FORSCHUNGSBERICHTE AUS DEM LEHRSTUHL FÜR REGELUNGSSYSTEME

TECHNISCHE UNIVERSITÄT KAISERSLAUTERN

Band 22

**Forschungsberichte aus dem Lehrstuhl
für Regelungssysteme
Technische Universität Kaiserslautern**

Band 22

Herausgeber:

Prof. Dr. Steven Liu

Min Wu

A Hybrid Physical and Data-driven Approach to Motion Prediction and Control in Human-Robot Collaboration

Logos Verlag Berlin

λογος

Forschungsberichte aus dem Lehrstuhl für Regelungssysteme
Technische Universität Kaiserslautern

Herausgegeben von
Univ.-Prof. Dr.-Ing. Steven Liu
Lehrstuhl für Regelungssysteme
Technische Universität Kaiserslautern
Erwin-Schrödinger-Str. 12/332
D-67663 Kaiserslautern
E-Mail: sliu@eit.uni-kl.de

Bibliographic information published by the Deutsche Nationalbibliothek

The Deutsche Nationalbibliothek lists this publication in the Deutsche Nationalbibliografie; detailed bibliographic data are available on the Internet at http://dnb.d-nb.de .

ISBN 978-3-8325-5484-2
ISSN 2190-7897

Logos Verlag Berlin GmbH
Georg-Knorr-Str. 4, Geb. 10,
12681 Berlin
Tel.: +49 (0)30 / 42 85 10 90
Fax: +49 (0)30 / 42 85 10 92
http://www.logos-verlag.de

Acknowledgment

This thesis presents the results of my research work at the Institute of Control Systems (Lehrstuhl für Regelungssysteme, LRS) in the Department of Electrical and Computer Engineering at the Technical University of Kaiserslautern.

First and foremost, I would like to express my great gratitude to my supervisor Prof. Dr.-Ing. Steven Liu. Over the last 13 years, he has always been an excellent supporter and encourager of my work from my undergraduate to doctoral study. He gives me a lot of free space to create my own ideas and provides valuable scientific guidance and discussions during my research journey.

Furthermore, my heartfelt thanks also go to Prof. Dr.-Ing. Uwe Hanebeck for his interest in my research and for joining the thesis committee as the second reviewer. His extensive expertise in machine learning and stochastic systems helps me a lot in improving my thesis. Additionally, I would like to thank Prof. Dr.-Ing. Norbert Wehn for joining the thesis committee as a chairman and for his invaluable discussions during the examination.

My time at LRS has been enjoyable and rewarding. My special thanks go to Dr.-Ing. Yanhao He, who always has an open ear to my questions and supports me a lot in experimental work. He is also an incredible friend, and I will never forget the great time we spent together. I would like to thank Prof. Dr.-Ing Daniel Görges for his support and wonderful discussion in the area of optimal control. I would like to thank all the other current and former colleagues at LRS for creating such an inspiring and cooperative atmosphere. Without loss of generality, I would like to mention M.Sc. Chen Cai, M.Sc. Xiang Chen, M.Sc. Sebastian Bard, M.Sc. Pedro Dos Santos, M.Sc. Marco Guerreiro, Dipl.-Ing. Adrian Herbst, M.Sc. Muhammad Ikhsan, M.Sc. Kashif Iqbal, M.Sc. Joe Ismail, M.Sc. Christoph Mark, M.Sc. Qingshan Pan, M.Sc. Tim Steiner, M.Sc. Jonas Ulmen, Dr.-Ing. Markus Bell, Dr.-Ing. Felix Berkel, Dr.-Ing. Sebastian Caba, M.Sc. Filipe Figueiredo, Prof. Dr.-Ing. Fabian Kennel, M.Sc. Markus Lepper, Dr.-Ing. Xiaohai Lin, Dr.-Ing. Tim Nagel, Dipl.-Ing. Peter Müller, Dr.-Ing. Sven Reimann, Dr.-Ing. Stefan Simon, Dr.-Ing. Alen Turnwald, Priv.Doz. Dr.-Ing. Christian Tuttas, Dr.-Ing. Yun Wan, Dr.-Ing. Hengyi Wang, Dr.-Ing. Jianfei Wang, MSc. Benjamin Watkins, Dr.-Ing. Wei Wu, Dr.-Ing. Zhuoqi Zeng and M.Sc. Yakun

I

Zhou. Moreover, I would like to thank the technicians Swen Becker and Thomas Janz and the secretaries Annegret Stabel and Jutta Lenhardt for their technical and administrative support.

My special thanks also go to colleagues in the research group wearHEALTH in the Department of Computer Science at the Technical University of Kaiserslautern. I would like to thank Dr.rer.nat. Bertram Taetz, Dr.-Ing. Markus Miezal, M.Sc. Michael Lorenz and Dr.-Ing. Gabriele Bleser for their collaboration and great support in the field of human motion capturing and analysis.

Last but not least, I would like to express my gratitude to my parents for their love and support throughout all the years. I am so lucky to have my wife Dr.rer.nat. Jingyun Chi by my side, being so understanding and supportive during the last years. Thank her for bringing two angels Anya and Anwen into the world, who are the most beautiful sunshine in my life.

Kaiserslautern, March 2022   *Min Wu*

Contents

List of Tables

List of Figures

Notation

Throughout the thesis, scalars are denoted by non-bold letters (e.g., a), matrices and vectors by bold letters (e.g., $\boldsymbol{a}$) and the real number set by double-struck letter $\mathbb{R}$.

Acronyms

ADP	Approximate Dynamic Programming
ANN	Adaptive Neural Network
ARE	Algebratic Riccati Euqation
CA	Constant Accelaration
CV	Constant Velocity
DMP	Dynamic Movement Primitives
DOF	Degree of Freedom
DP	Dynamic Programming
DS	Dynamic System
EMG	Electromyography
GMM	Gaussian Mixture Model
GP	Gaussian Process
HJB	Hamilton-Jacobian-Bellman
HMM	Hidden Markov Model
HRC	Human Robot Collaboration
IMU	Inertial Measurement Units
IOC	Inverse Optimal Control
IRL	Inverse Reinforcement Learning
KF	Kalman Filter
LQR	Linear Quadratic Regulator
LWR	Local Weighted Regression
MPC	Model Predictive Control

NRMSE	Normalized Root Mean Square Error
PE	Persistent Excited
PHRC	Physical Human Robot Collaboration
ProMPs	Probablistic Movement Primitives
RCC	Remote Center of Compliance
rECPD	rigid Extended Coherent Point Drift
RLS	Recursive Least Square
RL	Reinforcement Learning
RNN	Recurrent Neural Network
SONIG	Sparse Online Noisy Input Gaussain Process
TD	Temporal Difference
VPA	Value Function Approximation

Operators

$A \succ 0$	Positive definite
A^{-1}	Inverse of matrix A
$A^\dagger$	Pserdo inverse of matrix A
A^T	Transpose of matrix A
$\otimes$	Kronecker product
$\mathrm{diag}(A_1, ...)$	Block-diagonal matrix with $A_1...$
$\mathrm{rank}(A)$	Rank of matrix A

Chapter 2

δ	Measurement noise in KF
Φ	Regressor
Π	Parameter vector
Σ	Covariance matrix
ξ	Process noise in KF
a	Accleration
I	Identity matrix

P	Parameter in RLS
Q	Process variance in KF
R	Measurement variance in KF
S	Parameter in KF
v	Velocity
x	Position
$(\hat{\cdot})$	Estimated value
A	System matrix
B	Input matrix
C	Output matrix
K	Kalman gain
x	System state
$\mathbf{x}_0$	Initial state
y	System output
$J(\cdot)$	Cost function
t	Time
T_s	Sampling time

Chapter 3

$(\cdot)^*$	Referring to optimal value
$(\cdot)_*$	Referring to new incoming data
$(\cdot)_u$	Referring to to inducing point set
β	Cooficients of the basis functions in GP
Λ	Length scale in GP
μ	Expectation
ϕ	Feature function
Σ	Variance
σ	Measurement noise in GP
Θ	Parameter in IOC, weighting matrix in the cost function
θ	Parameter vector in IOC
ζ	Expert trajectory in IOC
a	Accleration
B	Variance of β

$\boldsymbol{b}$	Mean of $\boldsymbol{\beta}$
$\boldsymbol{h}$	Basis function set
$\boldsymbol{I}$	Identity matrix
$\boldsymbol{K}$	Corariance matrix of the traning set in GP
$\boldsymbol{k}$	Corariance function in GP
$\boldsymbol{m}$	Mean function in GP
$\boldsymbol{v}$	Velocity
$\boldsymbol{W}$	Weighting matrix
$\boldsymbol{X}$	GP input
$\boldsymbol{x}$	Position
$\boldsymbol{X}_+$	Incoming GP input
$\boldsymbol{Y}$	GP output
$(\hat{\cdot})$	Estimated value
g	Parameter in IOC, first-order partial derivate of the cost fuction
H	Parameter in IOC, second-order partial derivate of the cost fuction
K	Parameter in Online GP
L	Parameter in Online GP
u	System input
x	System state
$\mathbf{x}_0$	Initial state
$\mathcal{K}$	Control gain determined by IOC
σ_m	Signal variance in GP corariance function
σ_n	Output noise in GP
$J(\cdot)$	Cost function
p	Probability
t	Time
T_s	Sampling time

Chapter 4

$(\cdot)_\infty$	Referring to steady state
$(\cdot)_n$	Referring to nominal trajectory
α	Parameter in the canonical system

β	Parameter in the artificial congvergent behavior
$\boldsymbol{\kappa}$	Cross product of $\boldsymbol{n}$ and $\boldsymbol{n}_0$
$\boldsymbol{\Omega}$	Matrix form of Ω, containing multiple samples
$\boldsymbol{\omega}$	Weighting fuction between linear and nonlinear terms in DMP
$\boldsymbol{\Phi}$	Regressor
$\boldsymbol{\Psi}$	Matrix form of ψ, containing multiple samples
$\boldsymbol{a}$	Acceleration
$\boldsymbol{F}$	Matrix form of $\boldsymbol{f}$, containing multiple samples
$\boldsymbol{f}$	Nonlinear forcing term in DMP
$\boldsymbol{g}$	Target position
$\boldsymbol{I}$	Identity matrix
$\boldsymbol{k}$	Covariance function in GP
$\boldsymbol{K}, \boldsymbol{D}$	Gain Matrix in DMP
$\boldsymbol{m}$	Mean function in GP
$\boldsymbol{M}_c, \boldsymbol{K}_c, \boldsymbol{D}_c$	Control gains
$\boldsymbol{n}$	Unit vector, connecting intial-and target position
$\boldsymbol{n}_0$	Initial value of $\boldsymbol{n}$
$\boldsymbol{R}$	Rotation matrix
$\boldsymbol{S}$	Skew-symetric matrix
$\boldsymbol{v}$	Velocity
$\boldsymbol{v}_g$	Target velocity
$\boldsymbol{x}$	Position
$\boldsymbol{x}_0$	Inital position
γ	Parameter in the spatial scaling factor
$(\hat{\cdot})$	Estimated value
Ω	Weighting factor
ψ	Basis function
τ	Temporal scaling factor
θ	Dot product of $\boldsymbol{n}$ and $\boldsymbol{n}_0$
b, c	Parameters of the basis function
s	Phase variable

Chapter 6

$(\cdot)^0$	Referring to the world coordinate frame
$(\cdot)_H$	Referring to the human
$(\cdot)_o$	Referring to the object
$(\cdot)_R$	Referring to the robot
$(\cdot)_\Sigma$	Sum of the values
$\boldsymbol{\omega}$	Angular velocity
$\boldsymbol{\tau}$	Joint torques
$\boldsymbol{\tau}_{ext}$	External joint torques
$\boldsymbol{\theta}$	End-effector Rotation
$\boldsymbol{C}$	Centrifugal and Coriolis matrix
$\boldsymbol{G}$	Grasping matrix
$\boldsymbol{g}$	Gravitational torques
$\boldsymbol{h}$	Wrench
$\boldsymbol{I}$	Identity matrix
$\boldsymbol{J}$	Jacobian matrix
$\boldsymbol{M}$	Joint inertia matrix
$\boldsymbol{M}_d, \boldsymbol{K}_d, \boldsymbol{D}_d$	Desired impedance parameters
$\boldsymbol{p}$	Position
$\boldsymbol{q}$	Joint angle
$\boldsymbol{q}_d$	Joint angle reference
$\boldsymbol{R}$	Rotation matrix
$\boldsymbol{r}$	Relative position vector
$\boldsymbol{S}$	Skew-symetric matrix
$\boldsymbol{T}$	Homogeneous transformation matrix
$\boldsymbol{x}$	Robot end-effector pose in Cartesian space
$\boldsymbol{x}_d$	End-effector pose reference
$(\hat{\cdot})$	Estimated value
$\mathbf{Q}, \mathbf{R}$	Weighting matrices in the cost function
$\mathbf{X}$	State variable
K	Kinetic energy
P	Potential energy

$\tilde{q}$	Joint angle error
$\tilde{x}$	End-effector pose error
H_R	Hamilton function
r_R	Reward function
V_R	Value function

Chapter 7

$(\cdot)^*$	Referring to optimal value
$(\cdot)_H$	Referring to the human
$(\cdot)_R$	Referring to the robot
α_i	Discount factor of impedance parameters
$\mathcal{X}$	State feasible set
μ	Control policy
ϕ	Basis functions
τ	Joint torques
τ_{ff}	Feed-forward torque
τ_{imp}	Impedance torque
A	Continuous system matrix
A_d	Discrete system matrix
B	Continuous input matrix
B_d	Discrete input matrix
C	Centrifugal and Coriolis matrix
F	Transition function
f_{imp}	Impedance force
g	Gravitational torques
H, k	Paramter in joint constraints
I	Identity matrix
J	Jacobian matrix
l	Step size vector for TD-learning
M	Joint inertia matrix
M_d, K_d, D_d	Desired impedance parameters
p	Position

$\boldsymbol{P}, \boldsymbol{S}, \boldsymbol{\Phi}$	Parameter matrices in Q-learning
$\boldsymbol{p}_d$	Position reference
$\boldsymbol{u}$	System input
$\boldsymbol{W}$	Joint torque scaling factor
Δt	Sampling time
γ	Discount factor in the value function
$(\hat{\cdot})$	Estimated value
$\mathbf{Q}, \mathbf{R}$	Weighting matrices in the cost function
$\mathbf{X}$	State variable
$\mathcal{S}$	Vector of all elements in $\boldsymbol{S}$
τ'	Nominated joint torque
τ_{lim}	Joint torque limit
$\tilde{\boldsymbol{p}}$	Position error
e	TD-error
f_i	Interaction force
$f_{t,i}$	Interaction force threshold
Q	Q-function
r	Reward function
t	Time
V	Value function

Chapter 8

$(\cdot)_H$	Referring to the human
$(\cdot)_R$	Referring to the robot
$\boldsymbol{\beta}$	Parameter vector
$\boldsymbol{\Psi}$	Regressor
$\boldsymbol{\Sigma}_*$	Variance matrix in GP
$\boldsymbol{e}$	Estimation error
$\boldsymbol{g}$	Target position
$\boldsymbol{I}$	Identity matrix
$\boldsymbol{l}$	Position offset
$\boldsymbol{M}_d, \boldsymbol{K}_d, \boldsymbol{D}_d$	Desire impedance parameters

$\boldsymbol{M}_o$	Object inertia matrix
$\boldsymbol{P}$	Gain matrix
$\boldsymbol{R}$	Rotation matrix
$\boldsymbol{S}$	Parameter matrices in Q-learning
$\boldsymbol{T}_f$	Total travel time
$\boldsymbol{v}$	Velocity
$\boldsymbol{v}_g$	Target velocity
$\boldsymbol{x}$	Position
$\boldsymbol{x}_T$	Target position
$(\hat{\cdot})$	Estimated value
$\mathbf{Q}, \mathbf{R}$	Weighting matrices in the cost function
a_1, a_2	Parameters in the danger index
d	Danger index

1 Introduction

1.1 Motivation

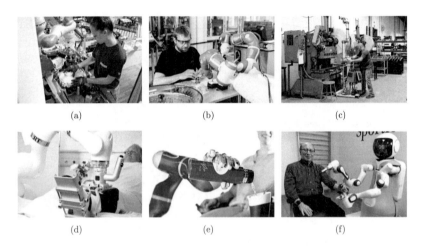

(a) (b) (c)

(d) (e) (f)

Figure 1.1: Application examples of cobots: (a) LBR iiwa, developed by KUKA, in
BMW production line, (b) Yumi, developed by ABB, in Elektro-Praga assembly line, (c) UR10, developed by Universal Robots, in GM production
line, (d) ROBERT, developed by KUKA, in rehabilitation of bedridden patients, (e) EDAN, developed by DLR, filling a drinking cup controlled from
the wheelchair, (f) GARMI, developed by MSRM, assistant for the elderly.

In recent years, increasing applications of collaborative robots (in short "cobots") have
been found not only in modern manufacturing systems (Fig. 1.1(a)-1.1(c)) but also in
daily personal services such as nursing care, household, and physiotherapy (Fig. 1.1(d)-
1.1(f)) [1]. According to the statistics from International Federation of Robotics, during

---

[1]Image source: KUKA, ABB, Universal Robot, DLR, MSRM,

the period from 2017 to 2019, the number of installed cobots in industry grew from $11,000$ to $18,000$ units[2]. The fast-growing demands for cobots have also led to increasing interest in the research community, which yields a new field: *human-robot collaboration* (HRC). As presented in [1], the number of publications per year on the topic of HRC grew from less than 20 to almost 800 from 1996 to 2015.

Despite the rapid growth of technologies and research in this area, today's HRC is still facing great challenges. In most industrial scenarios, although cobots work with humans in a shared workspace, their movements are sequential [2]. If the risk of collision arises, the robot completely stops until the human is out of the "danger zone". Very few applications can be found in which the robot adjusts its motion actively in real-time to the movement of the human partner. In the healthcare and service areas, most robots are still controlled by humans, either through teaching by hand or a teleoperation device. On the other hand, a recent survey[3] points out that from 2015-2018, the most developed research category in HRC is safety, which accounts for 64.2% of the identified papers. In summary, the technique and academic developments have achieved great success in solving the problem of "coexistence" between humans and robots in close proximity, but there is still a long way until "collaboration" can be realized.

What is a collaboration? Unfortunately, there is no unified definition of the term "collaboration". As reviewed in [4], different research communities such as robotics, human-machine interaction, cognitive science, multi-agent systems, etc., have their own domain-specific descriptions, and the boundaries between "collaboration", "cooperation", "coordination" and "joint action" are blurring. Nevertheless, it is commonly agreed that in a collaboration, each participant should meet the following fundamental criteria [5, 6, 7, 8]:

- representation of goals/tasks,
- monitoring and prediction of its own actions as well as the others',
- ability to interfere the individual and group behaviors towards goals.

To satisfy the "minimal" conditions listed above and achieve a true sense of human-robot collaboration, a "perception-analysis-interference" structure is suggested (Fig. 1.2) from the control engineering point of view, which has a similar formulation with classical feedback-control loop. Firstly, the perception module is responsible for measuring the states of humans which are not limited to the physical level (position, velocity, force) but also include cognitive features such as speech, emotion, gaze, etc. Wearable sensors have become extremely useful in providing accurate and reliable measurements of human activities [9]. Secondly, the analysis module interprets the measurements with appropriate models to recognize human intention, to predict human motion at a certain

---

[2]Souse: IFR World Robotics Report 2020

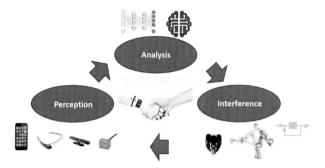

Figure 1.2: A perception-analysis-interference structure for human-robot collaboration

time in future, and to decide how the robot should react. Artificial intelligence and machine learning are the core technologies. Thirdly, the interference module follows the determined coordination strategies, generates local motion and control references, and executes in the physical environment. Control engineering plays an essential role in this process. The whole system should be able to run in real-time and adapt to the change of environments (including humans).

The structure has an interdisciplinary nature and includes many challenging problems. Although each component has been intensively studied by different research communities and yielded plenty of exciting results, there is still a fairly large gap between them. Very few examples can be found in which all the three fundamental elements are jointly considered. In the author's view, it is mainly due to the fact that systematic designs of HRC are still largely missing. Note that a systematic design does not simply mean the integration of various techniques into one system, which could be problematic since they have been developed from different theoretical backgrounds and under different practical conditions. More importantly, the design should pay particular attention to interconnections between the three fundamental components from both theoretical and technical perspectives. For instance, when choosing a modeling approach for the representation and prediction of human activities, it should be considered whether it is appropriate and applicable for robot control design, especially for the stability and feasibility examination.

Based on the perception-analysis-interference structure, the main goal of this thesis is to develop a novel trajectory planning and control framework for a safe, natural and effective human-robot collaboration. More specifically, the thesis focuses on (1) monitoring and predicting human hand motion in a collaborative manipulation task, (2) adjusting robot trajectories and forces online based on human motion predictions.

1.2 Scenarios studied in this thesis

The proposed framework is validated through two typical benchmark applications for HRC: object handover and cooperative object handling.

Object handover is a must-have skill for robots in both industrial manipulations and daily personal services. For instance, a robot co-worker should pass a tool to a human operator [10]. A service robot needs to bring drinks or medicine to a human patient [11]. To perform a handover as efficiently and fluently as a human is still a challenging task for robots. As comprehensively reviewed in [12], a handover requires "perception, prediction, action, learning, and adjustment by both agents".

This thesis concentrates on the "pre-handover" phase, in which the human and the robot move towards each other to get close enough to transfer the object. Precise control of the grasping motion/force will not be discussed in this thesis. The goal is to perform an on-the-fly handover, i.e., the robot should move simultaneously with the human. To enhance the flexibility, the initial pose and the object exchange location are not fixed. The key aspect for achieving this goal is online trajectory planning based on human motion prediction. For this purpose, modeling of human hand movements is needed. The main challenges are: (1) The model should cover the most characteristic features of human motion. Due to the high complexity of the human body dynamics, proper approximation methods are required. (2) The model should be computationally efficient for real-time implementations. (3) Because of the randomness of human motion, the model should be capable of dealing with uncertainties.

The second application studied in this thesis is collaborative object handling, in which the human and the robot jointly transport a rigid object. It is one of the most common benchmarks to investigate physical human-robot collaboration (pHRC). The main challenge is that the human and the robot have to agree on their movement directions and speeds [1]. One possible solution is to control the robot to behave compliantly when interacting with human. Under this concept, the robot only works passively under human guidance and results in a human-centered collaboration manner. It causes extra human effort and reduces adaptability and flexibility of the collaboration [13]. To overcome these drawbacks, the robot should be able to make proactive contributions based on human intention recognition and motion prediction. Furthermore, contact force is an important interaction modality in pHRC. It should be particularly considered and carefully handled in robot control design.

1.3 State-of-the-art technologies

1.3.1 Human modeling

As reviewed in [14], there are various techniques for modeling human behaviors, depending on different timescales and levels. This thesis mainly focuses on the motion level, which aims to represent relationships among physical variables such as velocity, acceleration, force, torque, etc. Cognitive modeling of human learning and coordination mechanisms is beyond the scope of this thesis.

Physical-based approaches

Conventional physical-based approaches use differential/difference equations derived based on physical laws to describe human motion dynamics. Such physical-based modeling can be categorized into kinematic and dynamic models. Kinematic models usually do not consider the forces and torques that produce the motion. The simplest examples are constant velocity (CV) and constant acceleration (CA) models. Both assume piecewise constant states (velocity or acceleration) with additive white noise [15]. One common application of such simple kinematic models is pedestrian motion prediction [16, 17, 18]. If they are suitable for describing the reaching motion of the human hand remains unclear and needs to be verified. A more widely used approach considers multiple degrees of freedom (DOF) of the human arm, which result in a series kinematic chain consisting of several joints and links [19]. Applications of such models, also known as skeleton-based models, require motion capture sensing systems to measure human body segments [20]. Commonly applied techniques include wearable Inertial Measurement Units (IMU) [21], optical motion capture systems [22] and fusion-based multi-sensor systems [23].

Dynamic models aim to describe the relationship between force and motion. Such models are usually needed in studying contact-reach scenarios, e.g., physical therapy [24], physical human-robot collaboration [25], and designing of humanoid robots [26] as well as exoskeletons[27]. One approach regards the human arm as a rigid body and derives the dynamic equations based on Lagrange formulation. The other considers the human arm as a series of elastic actuators and builds a mechanical impedance model with several springs and dampers. Both approaches require force/torque measurements for the identification of model parameters. A typical setup is that human grabs a haptic device or robot end-effector with integrated force/torque sensors and moves along a pre-specified path [28]. It strongly limits the implementation of manipulation tasks. Alternatively, Electromyography (EMG) signal can be used to measure muscle activities and estimate the change of force [29]. However, the data acquisition requires skin preparation and a long setup time.

Besides, another approach assumes that the human central nervous system synthesizes motion by minimizing a cost function consisting of physical variables. Oft-cited forms include minimum jerk model [30], minimum torque change model [31], minimum energy [32] model, etc. These "minimum-X" models were first introduced in the 1980 - the 1990s to fit point-to-point motion trajectories in free space. Nevertheless, several recent works have shown that with some extensions, e.g. online correction terms or parameter adaption [33, 34], these models can still achieve satisfying performance in human prediction for collaboration.

Physical-based approaches are suitable for representing physically well-understood systems and usually generalize well. However, they are still facing some difficulties in modeling human behaviors. Due to the high complexity of the human body, unknown dynamics still exist, which are not completely describable by physical knowledge. Hence, it is necessary to make approximations. An over-simplified model usually leads to poor performances, while a comprehensive model is computationally expensive and requires a large number of sensors to measure the corresponding physical quantities. Sensor drifts and signal distortions may cause errors in parameter identification. Another drawback is that physical models cannot handle uncertainties and randomnesses in human motion.

Data-driven approaches

Another category of human modeling concepts is known as data-driven approaches, which aim to find a functional relationship between pre-defined input and output variables based on a human motion data set. Since this functional relationship is mainly represented by probabilistic distributions, data-driven approaches are also named probabilistic approaches [35] and strongly associated with Bayesian theory and Gaussian distribution.

Commonly used data-driven models include Gaussian Mixture Model (GMM) [36], Gaussian Process (GP) [37], Hidden Markov Model (HMM)[38], etc. In recent years, following the breakthroughs in machine learning techniques, Recurrent Neural Networks (RNN) [39], Adaptive Neural Networks (ANN) [40] and deep learning [41] have also been applied in human motion prediction, utilizing their strength in learning complex and high dimensional systems. Another approach is Inverse Optimal Control (IOC) [42, 43]. The basic assumption is similar to the "minimum-X" models introduced above. The main difference is that the cost function is unknown and needs to be learned through observed human behaviors. Hence, this thesis also categorizes IOC into data-driven approaches.

Data-driven approaches require little prior knowledge on the process to be modeled and are flexible in describing a wide range of systems, especially for those that are not physically well-understood or too complex for deriving a physical model. Moreover, probabilistic representations enable data-driven models to deal with uncertainties. However, there are several critical problems. Since the data set is usually collected under certain labor conditions, it can only represent part of the system behaviors under constraints. As a result, the model may work well on training and test sets but loses its validity with unseen data. Moreover, due to a lack of prior knowledge, it remains unclear if the chosen variables are representative for describing the actual system dynamics. To overcome these problems, more data and variables need to be included in the data set, which dramatically increases its dimensions and computational complexity. In addition, the errors in data acquisition, e.g., sensor drift, noises, aliasing, etc., can also reduce the model accuracy.

Hybrid approaches

A new category of modeling approaches has drawn growing attention in recent years. The model consist of a physical-based term to describe the physically well-understood part of the system and a data-driven term to learn the rest of unknown dynamics. This category is known as "theory-guided data science"[44] or "hybrid machine learning"[45].

Also, in HRC, several related works can be found where hybrid approaches have been applied for modeling human motion dynamics. Typical examples are latent force model[46], dynamic systems [47], and dynamic movement primitives [48]. The ideas are similar: using an overly simplified mechanical model (e.g. a mass-spring-damper system) to cover some "basic" behaviors (e.g. a reaching movement from point A to B), then "refining" the model through machine learning techniques with human motion data. Some details will be discussed later in Chapter 4. Moreover, the model parameters can be further adapted online using various of learning algorithms to approximate time-varying human motion profiles during collaborations [49, 50, 51].

Hybrid approaches seem promising since they attempt to combine the strength of both physical-based and data-driven approaches. On the one hand, since the human motion profiles have been roughly described by the physical-based term, the data-driven term is only responsible for learning the physically not interpretable behaviors or the approximation errors. It can therefore enhance the learning effectiveness and reduce the size of the training set. On the other hand, hybrid approaches can achieve better generalizability than a purely data-driven model since the cause-effect relations between variables have been partially represented based on physical principles.

So far, investigations on the hybrid approaches for human modeling in HRC are still rare. In particular, there is a lack of well and systematically designed experimental validations.

1.3.2 Robot control

The robot control problem can be categorized into motion control in free space and control of the interaction with the environment [52]. The former aims to track a pre-defined reference motion trajectory, while the latter concentrates on regulating the contact force. In the two scenarios studied in this thesis, both motion and interaction control need to be addressed. Hence, it is desired to design a unified framework that can deal with both problems.

One possible approach is the hybrid position/force control. The principle is to decouple the control of end-effector position and contact forces into two orthogonal frames so that they can be executed independently from each other [52]. However, in HRC applications, this approach suffers from several difficulties. Firstly, the decoupling of motion and force control may not always be possible. Considering the collaborative object handling task, the robot should follow a reference path and simultaneously handle the contact force along the movement direction. Secondly, the constraint frames may change over time due to the varying human-robot contact geometry and needs to be updated online [53]. Thirdly, how to define a reasonable force reference value in HRC is still an open question, especially with consideration of subjective factors such as human comfortableness.

Another approach, namely the impedance control [54], has been commonly used in pHRC. The control goal is to achieve a desired relationship between robot end-effector motion and contact force, which is categorized through a mass-spring-damper system, known as mechanical impedance. Note that without contact force, the impedance control can also be used as a motion controller for compensating the position errors.

Conventional impedance control with pre-specified constant impedance parameters is not sufficient in the context of HRC. For instance, when tracking a reference path, the robot should maintain high impedance to suppress disturbances that perturb its end-effector from the desired trajectory. On the other hand, when the human partner intends to correct the robot motion, it should decrease the impedance to generate less resistant forces. Hence, modulating the desired impedance parameters to adjust robot compliance depending on task and interaction specifications has become an urgent research topic [55, 56]. Some previous works attempted to adapt the impedance parameters based on sensor feedback, e.g., robot end-effector velocity, contact force, etc.[57, 58, 59] To further enhance the performance, optimization-based approaches have been investigated. The impedance parameters are optimized by minimizing a task-dependent cost function subject to the robot and environment dynamics [60, 61, 62]. A fundamental challenge is that including the human as a part of the environment model will introduce a certain level of uncertainties or even unknown dynamics, which dramatically influence the solution of the model-based optimization problem. In recent years, learning-based impedance control has gained growing interest from HRC research. This approach extends the classical impedance control framework with machine learning techniques. The purpose is to develop an adaptive controller that is able to learn the environment model (including human), the impedance parameters, and/or the desired trajectories [63]. A detailed review of related research will be presented in Chapter 7. Since learning-based impedance control has already shown its advantages in cluttered and complex manipulation tasks, this thesis attempts to combine it with human motion prediction and design an adaptive motion generation and control framework.

1.3.3 Design examples

This section briefly reviews several design examples, in which human motion prediction, robot learning and control have been jointly considered in one framework.

The author in [64] develops an integrated framework combining human motion prediction with robot planning in real-time. The framework contains a data-driven multiple-predictor system that automatically identifies informative prediction features and combines the strengths of complementary prediction methods. Taking the prediction results as input, a feedback path planning algorithm is designed to adapt robot movements.

In [65], the author develops a *learn-collaborate-discover* architecture for cobots. In the *learn* module, the author proposed a geometric knowledge base for the robot, in which a complex manipulation task is represented by a series of kinematic constraints. This design significantly reduces the complexity of the learning procedure. The *collaborate* module is built upon the *learn* module for a human-in-the-loop execution of manipulation tasks with a particular concentration on the share-autonomy. The *discover* module aims to answer the question of how robots can learn from both observational and self-exploration when collaboration with humans.

The author in [66] addresses the problem of designing the behavior of cobots in dynamic, uncertain environments. A unique parallel planning and control architecture is presented, which contains a cognitive module for human behavior estimation and motion prediction, a goal-oriented long-term motion planner, and a safety-oriented short-term planer. All the planning and control algorithms are developed based on non-convex optimizations. Furthermore, the classification and modeling of the interaction modes in HRC are discussed.

The dissertation [67] focuses on the model-based design of a control and learning framework in physical HRC. In the learning part, the author proposed two modeling approaches of human behavior: (1) time-based Hidden Markov Models, which regards human motion as time-series, (2) an impedance-based Gaussian process, which takes the human arm impedance as priors. Both models are developed for online human motion prediction. In the control part, the author designs a stochastic optimal control approach in which uncertainties are not only considered in system models but also in costs. Moreover, the author discusses the roles of interaction forces and load distributions for designing anticipatory control schemes in physical HRC.

1.4 Objectives and Outline

As motivated by the state-of-the-art discussion, the overall objective of this thesis is to develop a novel methodological approach for human motion prediction and control in HRC. In particular, the thesis focuses on two relevant challenges:

1. development of an analytical, predictive and adaptive human motion model that is capable of efficient online human motion prediction under uncertainties, with specific considerations of its usefulness for the control design,

2. development of an adaptive control structure that is capable of handling unforeseen and time-varying changes in the environment (including the human partner) and enables robots to make proactive contributions in the collaboration instead of working passively as a follower under human guidance.

A hybrid design concept is presented throughout this thesis. On the one hand, it combines physical-based and data-driven approaches for modeling and predicting human motion. On the other hand, the state-of-the-art impedance control structure for cobots has been combined with learning-based techniques.

A graphical illustration of the thesis' structure and relations among different chapters is shown in Fig. 1.3. At the beginning of each chapter, a brief introduction is provided, including open questions, related work, and main contributions of this thesis.

Chapter 2 begins with classical approaches for human motion trajectory prediction based on recursive Bayesian estimation in combination with simple linear kinematic models. Such methods have been widely used for human motion tracking, which usually provides good short-term predictions. This chapter demonstrates two of the most commonly used kinematic models, constant acceleration model and minimum jerk model, with an extensive evaluation on the basis of two human motion datasets. The main focuses are

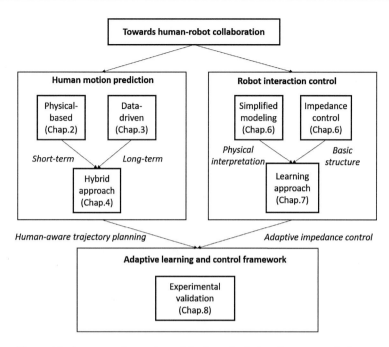

Figure 1.3: A structural overview of thesis and relations between each chapter.

analyses of the error rates with respect to prediction horizons and possible error sources.

Chapter 3 investigates data-driven approaches for human motion prediction. The purpose is to utilize their strength in representing complex systems to achieve long-term predictions. Motivated by the findings in the neuroscience literature, i.e., the human central nervous system works as an optimal feedback control system, the first part of this chapter aims to identify the objective function that the human tries to optimize, yielding an inverse optimal control problem. The second part focuses on modeling human motion through Gaussian process (GP), which belongs to non-parametric approaches. This work employs an online sparse GP regression method to overcome a well-known drawback of GP, namely the high computational complexity. Furthermore, the proposed method enables an adaptation of the GP model with new coming data. The evaluation of both methods includes not only the prediction error, but also their implementation complexity and generalizability.

Based on the findings in previous chapters, Chapter 4 introduces the concept of developing a hybrid physical-based and data-driven approach, i.e., using an overly simplified mechanical model to roughly describe the transitions of the observable physical states, then refining the model through machine learning techniques to achieve a more precise approximation of the system's dynamic. Within this concept, the chapter studies Dynamic Movement Primitives (DMP) and proposes special designs to overcome several limitations in the conventional DMP formulation. Results show that the proposed method outperforms all the other techniques that have been implemented so far in this work. Moreover, an extended DMP formulation to describe the interactive dynamics between humans and robots is presented at the end of this chapter.

Chapter 5 summarizes all the methods that have been studied in this work for human motion prediction and comprehensively discusses their scope of application, accuracy in different time scales, and implementation complexity. Afterward, an outlook on possible future research trends in human motion prediction based on hybrid approaches is stated.

Chapter 6 addresses robot interaction control in physical human-robot collaboration (PHRC). This work studies one typical benchmark application in which humans and robots jointly move a rigid object. Instead of a comprehensive kinematic and dynamic modeling of each agent with consideration of all degrees of freedoms, a more efficient approach is introduced, in which the compliance control behavior of both robots and humans is incorporated into the object dynamic. Accordingly, the basics of impedance control and its common forms are introduced. Lastly, the problem formulation of PHRC based on differential game theory is discussed.

Due to unknown parameters and high uncertainties in the human compliance control model, it is challenging to design robot control algorithms using model-based approaches. Reinforcement learning (RL) offers the possibility to learn an optimal control policy online through interaction with humans. Chapter 7 proposes a novel RL-based adaptive impedance control framework. Numerical simulations show that RL can achieve near-optimal performance without fully knowledge of the system dynamic. Moreover, possible extensions of the proposed method with consideration of constraints handling are discussed.

Chapter 8 presents the validation of the proposed learning and control methods by several human-robot collaboration experiments. Two typical benchmark applications, object-handover and object-handling are chosen so that both contact-free and physical interactions are included. To deal with various practical issues, several additional designs have been made to enhance the safety and adaptbility of the proposed framework. Comprehensive analyses and discussions on the experimental results are presented.

Chapter 9 provides conclusions of the main findings in this thesis and suggestions for future work.

1.5 Publications

Several articles on the motion prediction and control in HRC have been published during the doctoral studies. A chronological list of these articles together with their topic and their relation to this thesis is given below.

1. Min Wu, Bertram Taetz, Yanhao He, Gabriele Bleser, and Steven Liu: An Adaptive Learning and Control Framework based on Dynamic Movement Primitives with Application to Human-robot Handovers, *Robotics and Autonomous Systems* 148, 103935, 2022. (Chapter 4 and Chapter 8)

2. Min Wu, Yanhao He, and Steven Liu: Adaptive Impedance Control Based on Reinforcement Learning in a Human-Robot Collaboration Task With Human Reference Estimation, *International Journal of Mechanics and Control* , 21(01):21-32, 2020. (Chapter 7 and Chapter 8)

3. Min Wu, Bertram Taetz, Ernesto Dickel Saraiva, Gabriele Bleser, and Steven Liu: On-line motion prediction and adaptive control in human-robot handover tasks, *2019 IEEE International Conference on Advanced Robotics and its Social Impacts (ARSO)*, 1-6, 2019. (Chapter 3 and Chapter 8)

4. Min Wu, Yanhao He, and Steven Liu: Shared Impedance Control Based on Reinforcement Learning in a Human-Robot Collaboration Task, *International Conference on Robotics in Alpe-Adria Danube Region*, 95-103, 2019. (Chapter 7 and Chapter 8)

5. Min Wu, Yanhao He, and Steven Liu: Collaboration of multiple mobile manipulators with compliance based leader/follower approach, *2016 IEEE International Conference on Industrial Technology (ICIT)*, 48-53, 2016.

6. Yanhao He, Min Wu, and Steven Liu: A cooperative optimization strategy for distributed multi-robot manipulation with obstacle avoidance and internal performance maximization, *Mechatronics*, 76:12560, 2021.

7. Yanhao He, Min Wu, and Steven Liu: An optimisation-based distributed cooperative control for multi-robot manipulation with obstacle avoidance, *IFAC-PapersOnLine*, 53(2):9859-9864, 2020.

8. Yanhao He, Min Wu, and Steven Liu: Decentralised cooperative mobile manipulation with adaptive control parameters, *2018 IEEE Conference on Control Technology and Applications (CCTA)*, 82-87, 2018.

9. Henghua Shen, Ya-Jun Pan, Usman Ahmad, Steven Liu, Min Wu, and Yanhao He: Tracking Performance Evaluations on the Robust Teleoperative Control of Multiple Manipulators, *2019 IEEE 28th International Symposium on Industrial Electronics (ISIE)*, 1268-1273, 2019.

2 Human Motion Prediction based on Simple Kinematic Models

2.1 Introduction

This thesis investigates human motion prediction based on kinematic models. In particular, the human reaching motion has been studied, which is mainly generated by arm movements. Comprehensive modelling of the human arm with consideration of multiple degree-of-freedoms in shoulder, elbow and wrist is extremely complex [68] and is not suitable for control design. For simplification, this thesis focuses on describing human hand translational motion trajectories in Cartesian space without considering its dependency on the joint movements. This idea belongs to classical approaches in the area of human motion tracking. Commonly used models include constant velocity and constant acceleration models. Such models are usually combined with Bayesian filters (e.g. Kalman Filter) to deal with uncertainties [69]. Another approach assumes that the human central nervous system generates motion by minimizing an objective function over a time interval. Constraints such as initial and final conditions or via-points can be added to the optimization problem. The analog solutions usually result in a polynomial function of time.

These simple kinematic models are physics-based, easy to implement and do not occupy much computational resource when performed online. Hence, they have been still widely used and studied in various of human-robot-interaction scenarios. A recent study shows that in some cases such simple kinematic models can outperform even state-of-the-art neural network models [18]. This chapter presents a simulation study of two most commonly used kinematic models in literature, namely constant acceleration and minimum jerk model. Evaluation of both models is performed based on a human data set recorded by an optical motion capture system. Performances of both short-term and long-term predictions are analyzed and the possible source of error is discussed.

The remainder of the chapter is organized as follows. The scenarios studied in this thesis and the preparation of a human motion data set are briefly introduced in Section 2.2.

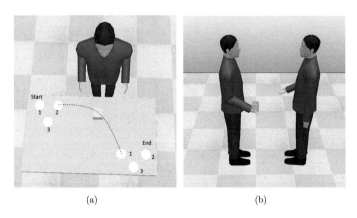

(a)   (b)

Figure 2.1: Graphical illustration of the scenarios for data collection: (a) pick and place
with different initial- and goal positions, (b)human-human object handovers

Section 2.3 describes the mathematical formulation of the two kinematic models and their
combinations with Bayesian filtering techniques. Section 2.4 presents the evaluation of
both models with measurement data and discusses the results. Section 2.5 summarizes
the pros and cons of the two models and gives references to further applications.

2.2 Data set preparation

For analysis and learning of human movements it is beneficial to build a human motion
data set, including trajectories performed by different participants under various experimental conditions. With the support of AG wearHEALTH at TU- Kaiserslautern [1], a
data set of human motion was built. The measurements were provided by OptiTrack
[2], a marker-based optical motion capture system contains 12 3D-range cameras with
maximal frame rate of 240 fps. The accuracy of motion tracking reaches 0.5 mm.

This thesis focuses on human reaching movements. Fig. 2.1 shows the two application
scenarios considered in the data collection. The left one is a typical "pick and place"
task with different initial and target positions. This scenario is usually seen in tabletop
manipulation tasks. For instance, in laboratories, the employees need to take and transfer
test tubes. Or in an assembling task, the workers collect and bring parts to different

---

[1]https://agw.cs.uni-kl.de/ Last visited: 01.02.2022
[2]https://optitrack.com/Last visited:01.02.2022

containers. During the data collection procedure, each participant was asked to pick up an object from one of the three starting positions and put it on one of the three target positions (see Fig. 2.1(a)). Each movement was repeated 20 times for 9 different starting- and end position combinations. Totally 5 participants took part in the measurement. According to actual needs, only upper body movements were recorded in a sampling frequency of 120 Hz.

The second scenario is an object handover between two human operators. The participants were asked to perform handovers naturally from different starting positions (see Fig. 2.1(b)). The meeting points and end poses were also not fixed. The hand movements of both giver and receiver, including transitional and rotational motion, were recorded under a sampling rate of 100 Hz. The aim is to study the joint activity and coordination strategies between humans. In comparison to the first scenario, the end-position of the reaching movement, namely the partner's hand position, is unknown and could vary according to the motion of both giver and receiver.

In the rest of this dissertation, all the studies on human motion trajectories were developed and validated based on this data set.

2.3 Mathematical formulations

2.3.1 Constant acceleration model

Many early approaches to human motion prediction use simple kinematic models, which take position, velocity and acceleration as states without considering forces that govern the motion. The constant acceleration model is one of the most commonly implemented models, especially for pedestrian motion prediction [15]. The key idea is that the acceleration between two samples is assumed to be constant, and the jerk (the derivative of acceleration) is regarded as white noise. Note that due to this assumption, the model has two limitations: (1) it works only with a short sampling time, (2) it is only suitable for short-term prediction.

The corresponding time-discrete state-space model is formulated as follows:

$$\underbrace{\begin{pmatrix} a(k+1) \\ v(k+1) \\ x(k+1) \end{pmatrix}}_{\mathbf{x}(k+1)} = \underbrace{\begin{pmatrix} I & 0 & 0 \\ T_s I & I & 0 \\ \frac{T_s^2}{2} I & T_s I & I \end{pmatrix}}_{\mathbf{A}} \underbrace{\begin{pmatrix} a(k) \\ v(k) \\ x(k) \end{pmatrix}}_{\mathbf{x}(k)} + \boldsymbol{\xi}(k), \quad \boldsymbol{\xi}(k) \sim \mathcal{N}(\mathbf{0}, \boldsymbol{Q}(k)) \qquad (2.1)$$

$$\underbrace{x(k)}_{\mathbf{y}(k)} = \underbrace{\begin{pmatrix} \mathbf{0} & \mathbf{0} & \mathbf{I} \end{pmatrix}}_{\mathbf{C}} \underbrace{\begin{pmatrix} a(k) \\ v(k) \\ x(k) \end{pmatrix}}_{\mathbf{x}(k)} + \boldsymbol{\delta}(k), \quad \boldsymbol{\delta}(k) \sim \mathcal{N}(\mathbf{0}, \boldsymbol{R}(k)), \tag{2.2}$$

where T_s is the sampling time, $\mathbf{x}, \mathbf{y}$ are the system state and output, $a, v, x \in \mathbb{R}^{3\times1}$ represent acceleration, velocity and position of the hand, $\boldsymbol{I} \in \mathbb{R}^{3\times3}$ is an identity matrix, $\mathbf{A}, \mathbf{C}$ are the system and output matrix, $\boldsymbol{\xi}, \boldsymbol{\delta}$ represent process and measurements noise with their variances $\boldsymbol{Q}$ and $\boldsymbol{R}$.

Considering uncertainties in the human kinematic motion as well as the measurement noise, the model is usually combined with Bayesian state estimation methods, e.g. Kalman Filter (KF), to achieve better tracking and prediction performance [69].

KF is an algorithm for exact Bayesian filtering with linear-Gaussian state-space models [70]. The algorithm contains two steps: a prediction step and a correction step. In the prediction step, the next state is estimated based on prior knowledge of the system, namely the dynamic model and its uncertainty described by the process noise. In the correction step, the estimated state is updated through a weighted average of the predicted value from the previous step and the actual noisy measurement. The weighting is determined by the Kalman gain. The computational details are summarized as follows:

1. Prediction step

$$\hat{\mathbf{x}}\left(k+1|k\right) = \mathbf{A}\mathbf{x}\left(k\right) \tag{2.3}$$

$$\hat{\boldsymbol{\Sigma}}\left(k+1|k\right) = \mathbf{A}\boldsymbol{\Sigma}\left(k\right)\mathbf{A}^T + \boldsymbol{Q}, \tag{2.4}$$

where $\boldsymbol{\Sigma}$ represents the error covariance matrix.

2. Correction step

$$\boldsymbol{S}\left(k\right) = \mathbf{C}\hat{\boldsymbol{\Sigma}}\left(k+1|k\right)\boldsymbol{C}^T + \boldsymbol{R}, \tag{2.5}$$

$$\mathbf{K}\left(k\right) = \hat{\boldsymbol{\Sigma}}\left(k+1|k\right)\mathbf{C}^T\boldsymbol{S}\left(k\right)^{-1}, \tag{2.6}$$

$$\mathbf{x}\left(k+1\right) = \hat{\mathbf{x}}\left(k+1|k\right) + \mathbf{K}\left(k\right)\left(\mathbf{y}\left(k\right) - \mathbf{C}\hat{\mathbf{x}}\left(k+1|k\right)\right) \tag{2.7}$$

$$\boldsymbol{\Sigma}\left(k+1\right) = \left(\boldsymbol{I} - \mathbf{K}\left(k\right)\boldsymbol{C}\right)\hat{\boldsymbol{\Sigma}}\left(k+1|k\right). \tag{2.8}$$

$\mathbf{K}$ is the Kalman gain matrix, which could be regarded as a weight between the covariance of the prior and the covariance of the measurement error. For example, if $\mathbf{K}$ is large, then more weight will be place on the correction term.

Principally KF only provides one-step prediction to have the optimal state estimation $\mathbf{x}\,(k+1)$. Nevertheless, it is also possible to achieve multi-step predictions by:

$$\hat{\mathbf{x}}\,(k+1+N) = \mathbf{A}^{N-1}\mathbf{x}\,(k+1)\,. \tag{2.9}$$

Note that since no measurement is available for $k+1, ...k+N$, the correction step cannot be performed from $k+2$, which makes the multi-step results less accuracy.

2.3.2 Minimum jerk model

As seen above, the constant acceleration model only focuses on describing the transition behaviour between two samples. Hence, it lacks "global" information of human motion trajectory, i.e. how does the motor control system generate human movements? The minimum jerk model, first introduced in [30], is one of the first answers to this question. It is developed for representing the human point-to-point reaching motion. The fundamental assumption is that human attends to minimize the time integral of the square of the jerk, when moving from an initial to a final position within a given time t_f. The motion trajectory is generated by solving the following constrained optimization problem:

$$J\,(\boldsymbol{x}) = \frac{1}{2}\int_0^{t_f}\|\dddot{\boldsymbol{x}}\,(t)\|^2 \tag{2.10}$$

$$\boldsymbol{x}^*\,(t) = \operatorname*{argmin}_{\boldsymbol{x}(t)}J \tag{2.11}$$

$$\text{s.t.}\quad \boldsymbol{x}\,(t_0) = \boldsymbol{x}_0 \tag{2.12}$$

$$\boldsymbol{x}\,(t_f) = \boldsymbol{x}_f \tag{2.13}$$

$$\ddot{\boldsymbol{x}}\,(t_0)\,,\ddot{\boldsymbol{x}}\,(t_f)\,,\dot{\boldsymbol{x}}\,(t_0)\,,\dot{\boldsymbol{x}}\,(t_f) = \mathbf{0}, \tag{2.14}$$

where $\boldsymbol{x} \in \mathbb{R}^{3\times 1}$ represents the human hand position measured in a fixed Cartesian coordinate, $\boldsymbol{x}_0$ and $\boldsymbol{x}_f$ are the initial- and goal positions. The solution results in a fifth-order polynomial time law for the position:

$$\boldsymbol{x}\,(t) = \boldsymbol{x}_0 + (\boldsymbol{x}_f - \boldsymbol{x}_0)\left(6\left(\frac{t}{t_f}\right)^5 - 15\left(\frac{t}{t_f}\right)^4 + 10\left(\frac{t}{t_f}\right)^3\right). \tag{2.15}$$

Taking the first derivative of the equation above, one will achieve the well-known bell-shaped velocity profile. Note that this form is also widely used in robot trajectory planning [52]. Moreover, the minimum-jerk model can also describe curved movements, e.g. for obstacle avoidance, by adding several via-points as constraints in the optimization problem [30].

Another advantage of the minimum jerk trajectory is its human preferment in context of a human-robot collaboration task. The user study [71] shows that in robot-human hand-over experiments, better results are achieved when using a minimum jerk model to generate robot trajectories than the conventional trapezoidal velocity profile. The participants feel more "safe" and "human-like" when the robot performs a minimum jerk end-effector motion.

According to Eq. (2.15), to predict human motion, it is necessary to know the goal position and the execution time. It could be critical in practice since both parameters are sometimes unknown or even time-variant. To deal with this issue, one can combine the minimum jerk model with efficient linear identification methods, which enables online estimation of unknown model parameters based on measurement data. Denoting $t_k = k \cdot T_s$, Eq. (2.15) is formulated as a linear model in the following sense:

$$\boldsymbol{x}\left(k\right) = \boldsymbol{\Phi}^T \boldsymbol{\Pi} + \boldsymbol{x}_0, \quad \boldsymbol{\Phi} = \begin{pmatrix} t_k^5 \boldsymbol{I} \\ t_k^4 \boldsymbol{I} \\ t_k^3 \boldsymbol{I} \end{pmatrix} \in \mathbb{R}^{9 \times 3} \tag{2.16}$$

where $\boldsymbol{I} \in \mathbb{R}^{3 \times 3}$ is an unit matrix, $\boldsymbol{\Pi} \in \mathbb{R}^{9 \times 1}$ is the parameter vector, which can be identified by Recursive Least Square (RLS) method [70]. If at t_k a measurement $\boldsymbol{x}\left(k\right)$ becomes available, then the estimated parameter vector $\hat{\boldsymbol{\Pi}}$ is iteratively updated by:

$$\hat{\boldsymbol{\Pi}}\left(k\right) = \hat{\boldsymbol{\Pi}}\left(k-1\right) + \frac{\boldsymbol{P}\left(k-1\right)\boldsymbol{\Phi}\left(k\right)}{\boldsymbol{I} + \boldsymbol{\Phi}^T\left(k\right)\boldsymbol{P}\left(k-1\right)\boldsymbol{\Phi}\left(k\right)} \cdot \left(\boldsymbol{x}\left(k\right) - \boldsymbol{\Phi}^T\left(k\right)\hat{\boldsymbol{\Pi}}\left(k-1\right) - \boldsymbol{x}_0\right),$$
$$\tag{2.17}$$

$$\boldsymbol{P}\left(k\right) = \boldsymbol{P}\left(k-1\right) - \frac{\boldsymbol{P}\left(k-1\right)\boldsymbol{\Phi}\left(k\right)}{\boldsymbol{I} + \boldsymbol{\Phi}^T\left(k\right)\boldsymbol{P}\left(k-1\right)\boldsymbol{\Phi}\left(k-1\right)} \cdot \boldsymbol{\Phi}^T\left(k\right)\boldsymbol{P}\left(k-1\right), \tag{2.18}$$

where the initial value $\boldsymbol{P}\left(0\right)$ is normally set large.

For fast convergence, the RLS method usually requires a good initial "guess" of the parameters. A common approach is to take a few measurements and use the ordinary Least Square method for an initial estimation of the parameters. Firstly a few samples $\boldsymbol{x}_{1:M} = \left(\boldsymbol{x}\left(1\right), \boldsymbol{x}\left(2\right), ..., \boldsymbol{x}\left(M\right)\right)^T$ are recorded with the regressor $\boldsymbol{\Phi}_{1:M} = \left(\boldsymbol{\Phi}\left(1\right), \boldsymbol{\Phi}\left(2\right), ..., \boldsymbol{\Phi}\left(M\right)\right)^T$ and written in a matrix form $\boldsymbol{x}_{1:M} = \boldsymbol{\Phi}_{1:M}\hat{\boldsymbol{\Pi}}_{\text{init}}$. Then the initial parameter estimation is performed by:

$$\hat{\boldsymbol{\Pi}}_{\text{init}} = \left(\boldsymbol{\Phi}_{1:M}^T \boldsymbol{\Phi}_{1:M}\right)^{-1} \boldsymbol{\Phi}_{1:M}^T \boldsymbol{x}_{1:M}. \tag{2.19}$$

For $k > M$ the parameter vector is updated by RLS. With the estimated parameters, a prediction of the future position with an arbitrary time horizon N within the execution time T is achieved by:

$$\hat{\boldsymbol{x}}\left(k+N\right) = \boldsymbol{\Phi}^T\left(k+N\right)\hat{\boldsymbol{\Pi}}\left(k\right) + \boldsymbol{x}_0. \tag{2.20}$$

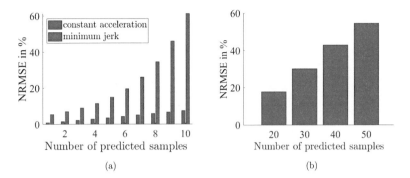

Figure 2.2: Performance of the short- and long-term online prediction based on constant acceleration (blue) and minimum jerk model (red) in a pick-and-place task.

2.4 Evaluation and discussion

In this section, the constant acceleration and minimum jerk model is evaluated and compared. Both table-top manipulation and handover scenarios are considered in order to study the difference in motion between a stand-alone and a cooperative task. As introduced before, both models are combined with recursive Bayesian estimation techniques (KF for constant acceleration model and RLS for minimum jerk model) to achieve an online prediction using incoming measurements. Note that in the minimum jerk model, the target position and total travelling time are unknown and need to be identified online.

Pick-and-place task

Firstly the human motion in a Pick-and-place task (Fig. 2.1(a)) is studied. 90 trajectories were randomly chosen from the data set to evaluate the model. The parameters of the Kalman Filter are chosen as follows:

$$Q = \begin{pmatrix} 0.1I^{3\times3} & & \\ & 0.01I^{3\times3} & \\ & & 0.001I^{3\times3} \end{pmatrix}, \quad R = 0.001I^{3\times3}.$$

Fig. 2.2 shows the results of a short- and long-term prediction based on the two models. Note that there is no standard definition of "short-" or "long-term". In this work, the

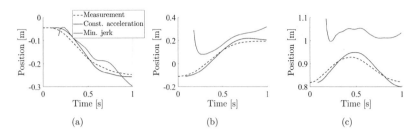

Figure 2.3: Prediction in one single demonstration with $N = 10$ based in constant acceleration (blue) and minimum jerk model (red) in a pick-and-place task:(a) $x-$dimension, (b) $y-$dimension, (c) $z-$dimension.

boundary is defined by 10% of the total trajectory length, which approximately results in 10 samples. The performance is evaluated by normalized root mean square error (NRMSE), which is calculated by:

$$\text{NRMSE} = \sqrt{\frac{\sum_{t=1}^{T} \|\hat{\boldsymbol{x}}(t) - \boldsymbol{x}(t)\|^2}{T}} \cdot \frac{1}{\|\max(\boldsymbol{x}) - \min(\boldsymbol{x})\|} \cdot 100\%, \qquad (2.21)$$

where $\hat{\boldsymbol{x}}, \boldsymbol{x}$ represent prediction and measurement respectively.

It can be seen in Fig. 2.2(a) that the constant acceleration model provides better results in comparison to the minimum jerk model for short-term prediction. The NRMSE of the prediction remains under 10 %. The overall performance of a long-term online prediction (prediction horizon ranges from 10 to 50) is shown in Fig. 2.2(b). Only results based on the constant acceleration model are plotted since the error in the minimum jerk model is too large. The diagram shows that the error increases rapidly along with the prediction horizon. As discussed before, the constant acceleration model was initially developed for tracking a moving target. The assumption of a "constant acceleration" is only valid in a short time range. Hence, it is not sufficient for long-term prediction.

Fig. 2.3 illustrates predicted trajectories with $N = 10$ in one single demonstration. Apparently, the constant acceleration model with KF follows the human trajectories well, while the minimum jerk model loses its validity.

The results show that the minimum jerk model is inaccurate for online prediction. One possible explanation is that the trajectory is computed by solving a global optimization problem. Locally the criterium may not always be satisfied. Another possibility is that besides the integral of jerk, the human motor control system aims to optimize other per-

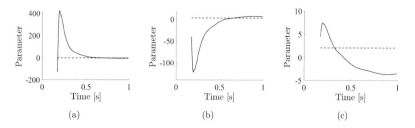

Figure 2.4: Convergence process of three parameters identified through RLS method in $x-$ dimension (dashed line: actual value. solid line: estimated value).

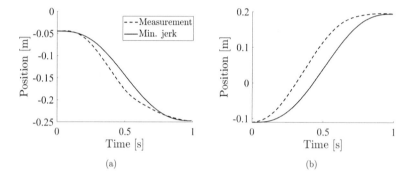

Figure 2.5: Offline trajectory reproduction using minimum jerk model with pre-specified parameters: (a) $x-$ dimension, (b) $y-$ dimension.

formance criteria so that the time law of the position cannot always be represented by a fifth-order polynomial. Consequently, these factors lead to errors in parameter identification (see Fig. 2.4), which strongly decreases the prediction performance. Another issue is that the convergence of the RLS method takes some time (see Fig. 2.4). Before the identified parameters converge, all the predictions are unreliable.

As presented in [30], the minimum jerk model has been developed for offline trajectory planning and human motion reproduction with pre-specified parameters. For validation, another test is performed, using the minimum jerk model to reproduce the human trajectories offline with a given target position and travelling time. The NRMSE results in 12.78 %, which is lower than the online case (compare to Fig. 2.2(a)). Fig. 2.5 shows the reproduced trajectories in one single demonstration. The performance is better than the

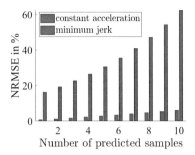

Figure 2.6: Performance of the short-term online prediction based on constant acceleration (blue) and minimum jerk model (red) in a handover task.

results showed in Fig.2.3.

Note that a particular characteristic of the table-top manipulation task is that in $z-$
direction, the initial and target positions are identical. Under this condition, the solution
of the minimum jerk problem in Eq. (2.10) is $x^*(t) = x_0, \forall t > 0$, which means the position
remains unchanged. To deal with this problem, one can add some via-points (e.g. the
maximum height) and formulate them as constraints. However, in this case, the solution
is not the same fifth-order polynomial as before.

Handover task

In an object handover task, the motions of both giver and receiver are correlated to
each other. It is interesting to study how large this correlation will influence the validity
of the model. For this purpose, the same kinematic models are used to predict human
motion in handovers. If the prediction error remains at the same level as before, one can
conclude that the previous models are valid in such a collaborative task as well.

As shown in Fig. 2.6, for constant acceleration model the prediction error has no significant difference in comparison to the previous scenario (Fig. 2.3). Thus, the zero-jerk
assumption is still valid, i.e., humans also intend to generate smooth motion in a collaborative task. On the other hand, the minimum jerk model provides a slightly worse
result, mainly due to the slow convergence and error in the parameter identification. In
a handover task, the time-varying handover location and duration bring more difficulties
in the identification process.

2.5 Summary

In this chapter, two classical kinematic models were introduced and implemented to predict human hand movements in Cartesian coordinate system. Simulation results show that the constant acceleration model combined with Kalman filter is suitable for short-term online human motion prediction. The mathematical formulation is simple, and no identification of the model parameter is required. The disadvantage is that it cannot achieve satisfied long-term prediction.

In comparison, the minimum jerk model is fit for offline planning/reproduction of the motion trajectory. The time law of the position results in a fifth-order polynomial with the target and duration as parameters. However, results in this work show that if these parameters are unknown, a simultaneous state and parameter estimation with this model is problematic. One possible explanation is that the fundamental assumption of "minimum jerk" was made from a global planning point of view. Locally this condition may not be always satisfied. Hence, with an incomplete trajectory observation the identification of model parameters cannot work properly. Another possible factor is there exists other criteria when human performs trajectory planning.

A special practical issue is that the mathematical formulation of the minimum jerk model loses validity when describing a trajectory in which the initial and target positions are close to each other. In this case, extra constraints such as via-points should be added to the optimization problem.

Despite several limitations in human motion prediction, these two simple kinematic models can be regarded as a standard baseline to develop advanced motion prediction and control methods. There exist various extensions and further applications of both models. For instance, the constant acceleration model can be combined with Model Predictive Control (MPC) to predict human motion and generate optimal robot trajectories in a dynamic environment to avoid collision [72]. Furthermore, due to its high tracking and short-term prediction performance, the constant acceleration model is used in this thesis to determine the velocity and acceleration value of both human hand and robot end-effector when only position measurement is available.

The minimum jerk model is still one of the most widely used movement patterns to generate a smooth trajectory in Cartesian space. As an example, authors in [73] used the minimum jerk model to generate a Cartesian reaching motion for a multi-degree-of-freedom robot arm, then transformed it into joint space via an inverse kinematic solver. Another interesting extension is an online minimum jerk trajectory generation algorithm proposed in [74]. The method was specially designed to deal with changes in the target position.

3 Data-driven Approaches for Human Motion Prediction

3.1 Introduction

As discussed in the previous chapter, computationally effective simple kinematic models are not sufficient to describe human reaching movement and suffer from problems in long-term prediction. A possible solution is to develop more comprehensive physical models, considering additional features (e.g., joint movements in shoulder, elbow, wrist). However, this will result in highly complex models, and the identification of model parameters is difficult. This chapter switches to another area, namely data-driven approaches for representing and predicting human motion.

As introduced in the first chapter, the main task in this thesis is to develop robot planning and control strategies to enhance the safety and efficiency of the collaboration. Hence, this chapter mainly focuses on the "computational level modelling" of human behavior [14], especially the objective function that syntheses human motion and the probabilistic description of trajectories.

The first part of this chapter introduces the inverse optimal control (IOC) method. It follows the assumption that "the human central nervous system works as an optimal feedback control system, in which the feedback gains are optimized based on some performance indexes" [75]. As shown in the previous chapter, the jerk is one possible candidate for such performance indexes. A more advanced concept would be combining multiple performance criteria in one objective function with different weightings and identifying these weightings through human demonstrations. The purpose of IOC is to recover the cost/reward function from observed human motion trajectories based on machine learning techniques. [76]. There exist many previous works using IOC to analyze the mechanism of the human motor system, including arm reaching motion [43, 77], locomotion [78, 79] or driving behavior [80, 81]. However, most of these works mainly focus on the explanation of the human motion planning synthesis rather than real-time motion prediction for control purposes. Authors in [42] proposed one of the first frameworks of predicting human motion based on IOC in a collaboration task. Even though their method outperforms other baseline methods, 23 DOFs of the human kinematic model are considered, making the computational load very high.

In this chapter, a two-phase framework by combining IOC with a simple kinematic model and linear quadratic regulator (LQR) is presented, which aims to predict human hand reaching movements. In the first phase, the IOC problem is solved based on the algorithm proposed in [76], which is computationally effective and suitable for continuous domains. In the second phase, an LQR problem with the identified cost function is solved, then the optimized feedback control gain is substituted into the system model to make predictions.

The second part of this chapter works directly on the motion trajectories and learns a statistical pattern of human motion behaviors with Gaussian Process (GP) regression. Motivated by its effectiveness in the representation of nonlinear dynamics with small data sets [82, 83], Gaussian Process regression is studied with special consideration of its online computational efficiency and adaptability to new incoming data.

The rest of this chapter is organized as follows: Section 3.2 outlines the concept, derivation, implementation and evaluation of the inverse optimal control method. Section 3.3 presents an online sparse Gaussian Process regression algorithm with an extensive evaluation on prediction error, scalability and generalizability. At last, Section 3.4 summarized the important findings in this work.

3.2 Inverse optimal control

3.2.1 Problem formulation

An optimal control problem aims to determine control inputs that make a dynamic system satisfy the physical constraints and at the same time minimize (maximize) some performance criteria measured by a cost (reward) function [84]. In contrast, inverse optimal control (IOC), also known as inverse reinforce learning (IRL), has the object to recover an unknown cost/reward function from control and state trajectories assumed to be optimal [76].

Considering the following time invariant dynamic system:

$$\mathbf{x}\left(k+1\right) = \boldsymbol{f}\left(\mathbf{x}\left(k\right), \mathbf{u}\left(k\right)\right)$$
$$\mathbf{x}\left(1\right) = \mathbf{x}_0, \tag{3.1}$$

where $\mathbf{x} \in \mathbb{R}^{n \times 1}$, $\mathbf{u} \in \mathbb{R}^{m \times 1}$ represent states and inputs, $\mathbf{x}_0$ is the initial state. Assuming that there exists a control input which minimizes a cost function $J(\mathbf{x}, \mathbf{u})$, namely:

$$\mathbf{u}^* = \arg\min_{\mathbf{u}} J(\mathbf{x}, \mathbf{u}), \tag{3.2}$$

The task is to determine $J(\mathbf{x}, \mathbf{u})$ with several optimized state and input trajectories with K−samples, denoted as:

$$\mathbf{x}^* = \left(\mathbf{x}^{*T}(1), \; ..., \; \mathbf{x}^{*T}(K)\right) \in \mathbb{R}^{n \times K}$$

$$\mathbf{u}^* = \left(\mathbf{u}^{*T}(1), \; ..., \; \mathbf{u}^{*T}(K)\right) \in \mathbb{R}^{m \times K}. \tag{3.3}$$

The cost function J is parameterized as a linear combination of M given features $\boldsymbol{\phi}(\mathbf{x}, \mathbf{u}) \in \mathbb{R}^{M \times 1}$, namely:

$$J(\mathbf{x}, \mathbf{u}) = \frac{1}{2} \sum_{k=1}^{K} \boldsymbol{\theta}^T \boldsymbol{\phi}(\mathbf{x}(k), \mathbf{u}(k)),$$

$$\text{s.t. } \mathbf{x}(k+1) = \boldsymbol{f}(\mathbf{x}(k), \mathbf{u}(k))$$

$$\mathbf{x}(1) = \mathbf{x}_0, \tag{3.4}$$

The IOC problem now becomes a parameter identification problem, namely determining parameters $\boldsymbol{\theta}$ of the cost function J, which is minimized by $\mathbf{u}^*$ [80].

3.2.2 Solution based on the principle of maximum entropy

As discussed in [85], a well-known problem of the IOC is ambiguity. Since different cost functions may have the same optimal policy, the learned cost function may contain a stochastic preference for one of them. Moreover, the observed trajectories are rarely perfectly optimal and contain "noise".[1] To deal with these problems, authors in [85] modelled the control policy as a probabilistic distribution and employ the principle of maximum entropy. Within this concept, the "best" model (distribution) is the one with largest entropy, which means it has no preference for any possible solution.

Denoting $\boldsymbol{\zeta} = (\mathbf{x}, \mathbf{u})$, based on the principle of maximum entropy, the probability of the optimal control input and state trajectories (expert trajectories) $\boldsymbol{\zeta}^* = (\mathbf{x}^*, \mathbf{u}^*)$ is proportional to the exponential of the cost encountered along the whole trajectory. It yields:

$$p(\boldsymbol{\zeta}^* | \mathbf{x}_0) = \frac{\exp(-J(\boldsymbol{\zeta}^*))}{\int \exp(-J(\boldsymbol{\zeta})) d\boldsymbol{\zeta}}, \tag{3.5}$$

---

[1]Here, the definition of "noise" is not limited to measurement noise. It also contains the deviation from the actual optimal solution.

where the denominator is usually named partition function and contains all the possible trajectories that satisfy Eq. (3.1).

Recalling Eq. (3.4), the inverse optimal problem is to find the optimal estimation of $\boldsymbol{\theta}$ so that:

$$\boldsymbol{\theta}^* = \arg\max_{\boldsymbol{\theta}} p\left(\boldsymbol{\zeta}^* | \mathbf{x}_0, \boldsymbol{\theta}\right). \tag{3.6}$$

Solving the above optimization problem requires computing the partition function, i.e., integral over all possible trajectories that satisfy the system dynamic equation, which is impossible in practice. To deal with this issue, authors in [76] proposed an approximate representation. The idea is explained briefly in the following and described in more detail in Appendix A. The key assumption is that "the expert (human) performs a local optimization when choosing the control input $\mathbf{u}$, rather than a global planning". Hence, the cost function can be approximated by a quadratic Taylor series expansion around $\mathbf{u}$. Under a further assumption that the distribution of $\boldsymbol{\zeta}$ is Gaussian, the probability described in Eq. (3.5) is simplified as:

$$p(\boldsymbol{\zeta}^* | \mathbf{x}_0) \approx \exp\left(-\frac{1}{2}\mathbf{g}^T\mathbf{H}^{-1}\mathbf{g}\right) \det(\mathbf{H})^{\frac{1}{2}}(2\pi)^{-\frac{mK}{2}}, \tag{3.7}$$

where $\mathbf{g} \in \mathbb{R}^{mK}$ and $\mathbf{H} \in \mathbb{R}^{mK \times nK}$ represent the first- and second order partial derivative of J with respect to $\boldsymbol{\zeta}^*$. They are calculated by:

$$\mathbf{g} = \frac{\partial J}{\partial \mathbf{u}^*} + \frac{\partial \mathbf{x}^{*}}{\partial \mathbf{u}^*}^T \frac{\partial J}{\partial \mathbf{x}^*}, \tag{3.8}$$

$$\mathbf{H} = \frac{\partial^2 J}{\partial \mathbf{u}^{*2}} + \frac{\partial \mathbf{x}^{*}}{\partial \mathbf{u}^*}^T \frac{\partial^2 J}{\partial \mathbf{x}^{*2}} \frac{\partial \mathbf{x}^*}{\partial \mathbf{u}^*}. \tag{3.9}$$

Based on the above simplification of $p(\boldsymbol{\zeta}^* | \mathbf{x}_0)$, the optimization defined in Eq. (3.6) is reformulated as:

$$
\begin{aligned}
\boldsymbol{\theta}^* &= \underset{\boldsymbol{\theta}}{\operatorname{argmax}}\, p(\boldsymbol{\zeta}^* | \mathbf{x}_0) \\
&\Leftrightarrow \underset{\boldsymbol{\theta}}{\operatorname{argmax}}\, \ln p(\boldsymbol{\zeta}^* | \mathbf{x}_0) \\
&\approx \underset{\boldsymbol{\theta}}{\operatorname{argmax}} \left(-\frac{1}{2}\mathbf{g}^T\mathbf{H}^{-1}\mathbf{g} + \frac{1}{2}\ln\det(\mathbf{H}) - \frac{mK}{2}\ln 2\pi \right) \\
&\Leftrightarrow \underset{\boldsymbol{\theta}}{\operatorname{argmin}} \left(\frac{1}{2}\mathbf{g}^T\mathbf{H}^{-1}\mathbf{g} - \frac{1}{2}\ln\det(\mathbf{H}) + \frac{mK}{2}\ln 2\pi \right).
\end{aligned} \tag{3.10}
$$

Note that the optimal variable $\boldsymbol{\theta}^*$ and the observed expert trajectories $\boldsymbol{\zeta}^*$ are all included in $\mathbf{g}$ and $\mathbf{H}$. More details and related derivations are given in Appendix A.

3.2.3 Implementation

In this thesis, IOC is used to predict human hand reaching movement with known initial and final positions. The implementation is performed based on the recorded data in the table-top manipulation task (Fig. 2.1(a)), which is randomly divided into a training and a test set. Firstly, trajectories in the training set are used to determine the optimal variable $\boldsymbol{\theta}^*$ as well as the cost function J. Then the optimization problem defined in Eq. (3.4) is solved using LQR to compute the optimal control gain. The closed-loop dynamic is used as the model for online prediction.

In order to simplify the computation and analysis, following assumptions are made:

- The state and input vectors are defined by

$$\mathbf{x} = \begin{pmatrix} \boldsymbol{x} - \boldsymbol{x}_T \\ \boldsymbol{v} \end{pmatrix} \in \mathbb{R}^{6\times1}, \quad \mathbf{u} = \boldsymbol{a} \in \mathbb{R}^{3\times1}, \tag{3.11}$$

where $\boldsymbol{x}, \boldsymbol{v}, \boldsymbol{a}$ represent human hand position, velocity and acceleration, $\boldsymbol{x}_T$ is the target position.

- The dynamic system is linear, namely:

$$\mathbf{x}(k+1) = \mathbf{A}\mathbf{x}(k) + \mathbf{B}\mathbf{u}(k), \quad \mathbf{A} = \begin{pmatrix} \boldsymbol{I} & T_s\boldsymbol{I} \\ \boldsymbol{0} & \boldsymbol{I} \end{pmatrix}, \quad \mathbf{B} = \begin{pmatrix} \frac{T_s^2}{2}\boldsymbol{I} \\ T_s\boldsymbol{I} \end{pmatrix} \tag{3.12}$$

where $\boldsymbol{I}^{3\times3}$ is an unit matrix, T_s represents the sampling time. Note that the mathematical formulation is exactly the same as constant acceleration model. The difference is that the acceleration is regarded here as system input rather than a state variable.

- The features that construct the cost function J are in a quadratic form, namely:

$$\boldsymbol{\phi}(k) = \begin{pmatrix} \mathbf{x}(k) \otimes \mathbf{x}(k) \\ \mathbf{u}(k) \otimes \mathbf{u}(k) \end{pmatrix} \in \mathbb{R}^{9\times1}. \tag{3.13}$$

Then the cost function is formulated as:

$$J = \frac{1}{2}\sum_{k=1}^{K}\left((\boldsymbol{x}(k) - \boldsymbol{x}_T)^T \boldsymbol{\Theta}_1 (\boldsymbol{x}(k) - \boldsymbol{x}_T) + \boldsymbol{v}(k)^T \boldsymbol{\Theta}_2 \boldsymbol{v}(k) + \mathbf{u}(k)^T \boldsymbol{\Theta}_3 \mathbf{u}(k) \right), \tag{3.14}$$

where $\boldsymbol{\Theta}_1 = \mathrm{diag}(\theta_1, \theta_2, \theta_3), \boldsymbol{\Theta}_2 = \mathrm{diag}(\theta_4, \theta_5, \theta_6), \boldsymbol{\Theta}_3 = \mathrm{diag}(\theta_7, \theta_8, \theta_9)$ are positive definite weighting matrices. From the physical point of view, the first term

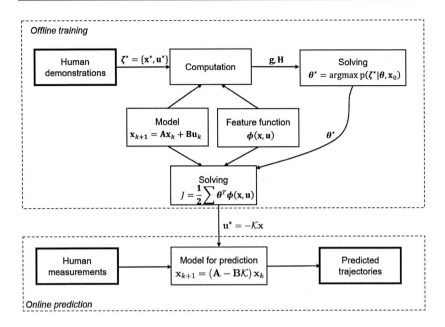

Figure 3.1: Flowchart of human motion prediction based on IOC

can be regarded as potential energy between the actual and target positions. The second term represents kinetic energy. The last term relies on the energy that generalizes the motion.

Fig. 3.1 illustrates the procedure of human motion prediction based on IOC. The process includes two phases. The offline training phase aims to solve the inverse optimization problem defined in Eq. (3.10) to get the best estimation of the parameter set $\boldsymbol{\theta}^*$. The components $\mathbf{g}$ and $\mathbf{H}$ are computed based on observed state and input trajectories, the system model, and the chosen features (see Appendix A). The problem is solved by a nonlinear programming solver *fmincon* in *Matlab* with constraint $\theta_i > 0, \forall i$.

In the next step, the original optimization problem defined in Eq.(3.4) with identified cost function is solved through linear quadratic programming. The solution results in an optimal state-feedback controller:

$$\mathbf{u}^* = -\mathcal{K}\mathbf{x}, \tag{3.15}$$

so that the closed-loop dynamic is described by the following equation:

$$\mathbf{x}(k+1) = (\mathbf{A} - \mathbf{B}\mathcal{K})\mathbf{x}(k). \tag{3.16}$$

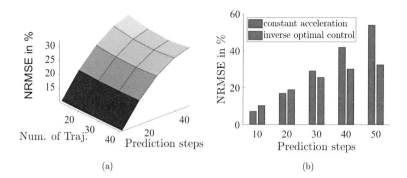

(a)   (b)

Figure 3.2: Performance of long-term motion prediction based on IOC: (a) performance over size of training set and prediction steps, (b) comparison with constant acceleration model.

For online prediction, at each sampling point k, the equation above is used as a model. Taking the current measurement $\mathbf{x}(k)$ into the equation and computing for N times provides a prediction of states $\hat{\mathbf{x}}(k+1),...\hat{\mathbf{x}}(k+N)$.

3.2.4 Evaluation and discussion

The study focuses on long-term human motion prediction with prediction steps N ranging from 10 to 50. The overall performance depending on the training set size as well as the prediction horizons, is plotted in Fig. 3.2(a). It can be seen that the number of training trajectories has no significant influence on the prediction error. Hence, a large data set is not necessary for training. Comparing to the constant acceleration model, the increasing rate of error over the prediction horizon is lower.

Fig 3.2(b) illustrates the performance in long-term prediction compared to the constant acceleration model. In the cases of $N = 10$ and 20, the error is even higher. As of $N = 30$ the advantage of IOC is notable. Especially for $N = 50$ the error is reduced by around 20%.

The identified weightings including their mean μ and standard deviation σ are listed in Table 3.1. The position-related term has the largest weighting, which is reasonable in a reaching movement. Specifically, the identified $z-$ component (θ_3) is near zero, which also matches to the table-top task since the initial and final positions are at the same

θ_i	μ	σ
1	59.5306	75.1150
2	153.6489	254.7109
3	0.0047	0.0219
4	0.2553	0.8958
5	2.1367	5.5651
6	0.0025	0.0147
7	0.9958	0.8884
8	1.2513	1.0930
9	1.1170	0.8221

Table 3.1: Identified parameters in the cost function in Eq.(3.14). 1-3: position relied term, 4-6: velocity relied term, 7-9: acceleration relied term.

height. Weightings between velocity and acceleration related terms have no significant difference. However, it should be pointed out that uncertainties in the identified parameters are extremely high (see the value of σ in Table 3.1). Similar phenomenon is observed in one related work [80], in which IOC is implemented to identify driver's habits. The authors indicate that the identified parameters vary from person to person. Hence, they argue that the IOC method is sufficient to study individual human behaviour. In other words, the identified cost function is not generalizable.

The motivation of implementing IOC is to investigate further how the motor control system generates human movements and utilize this knowledge to achieve more precise long-term motion prediction. The results show a certain degree of performance improvement in caparison with a simple kinematic model. However, in this work, the following disadvantages and limitations are discovered:

- The parameterization of the cost function as a linear combination of quadratic feature functions limits the class of optimization problems that the cost function can describe.

- The features defined in the cost function must be measurable/observable to gather the expert trajectories for solving the IOC problem. Moreover, a full trajectory observation is required.

- The whole procedure needs to solve two optimization problems. The choice and configuration of numerical solvers may influence the results.

- There exist significant uncertainties in the identified parameters.

- The algorithm used in this work can only learn an unconstrained optimization problem.

- The learned cost function is assumed to be time invariant.

Nevertheless, the IOC method helps to achieve a further understanding of human motion control strategies. The method can be extended by combining it with an online trajectory planning algorithm to generate human-aware robot motions [86]. Several recent work works provide promising results in dealing with some of the above-listed problems. For instance, authors in [87] provided an IOC algorithm that works on incomplete trajectory observations, which is computationally effective and can adapt to time-varying objective functions. Authors in [88] developed an IOC approach to learn both cost function parameters and constraints in a human manipulation task.

3.3 Gaussian Process Regression

This section aims to use data-driven approach to fit the human hand motion trajectories. Defining:

$$\mathbf{x} = \begin{pmatrix} \boldsymbol{x}(k) \\ \boldsymbol{v}(k) \\ t_k \end{pmatrix} \in \mathbb{R}^{7\times1}, \quad \mathbf{y} = \begin{pmatrix} \boldsymbol{x}(k+1) \\ \boldsymbol{v}(k+1) \end{pmatrix} \in \mathbb{R}^{6\times1}, \tag{3.17}$$

where $\boldsymbol{x}(k), \boldsymbol{v}(k) \in \mathbb{R}^{3\times1}$ are the human hand position and velocity, t_k represents time.

Assuming that the following functional relationship holds:

$$\mathbf{y} = \boldsymbol{f}(\mathbf{x}) + \boldsymbol{\sigma}, \qquad \boldsymbol{\sigma} \sim \mathcal{N}\left(0, \sigma_n^2 \boldsymbol{I}\right), \tag{3.18}$$

where $\boldsymbol{\sigma}$ is the measurement noise. The purpose is to determine the function $\boldsymbol{f}(\mathbf{x})$, which describes how the state (here: human position and velocity) evolves over time.

3.3.1 Regular form

Gaussian Process (GP) regression is a Bayesian machine learning approach, which provides a non-parametric and probabilistic representation of various systems. GP is defined as "a collection of random variables, any finite number of which have a joint Gaussian distribution" [89]. A function modeled by GP is denoted as follows:

$$f(\mathbf{x}) \sim \mathcal{GP}\left(m(\mathbf{x}), k(\mathbf{x}, \mathbf{x}')\right), \tag{3.19}$$

where m, k represent the mean- and covariance functions respectively, $\mathbf{x}'$ represents the state variable that differs from $\mathbf{x}$.

In literature, GP is usually categorized as non-parametric regression since it does not necessarily require any assumption on the parametric form of the function $f(\mathbf{x})$, such as type of the function, system order and linearity. It would reduce the danger that $f(\mathbf{x})$ is not well-modelled by the previously chosen class of functions. The only assumption is that the distribution of the functions is Gaussian, which takes advantage of the Gaussian distribution characteristics and simplifies the computation.

GP is entirely specified by its mean and covariance functions. For simplification, a zero-mean is often assumed. The covariance function can be represented by different kernel functions, for instance, the squared exponential function:

$$k(\mathbf{x}, \mathbf{x}') = \sigma_m^2 \exp\left(-\frac{1}{2}(\mathbf{x} - \mathbf{x}')^T \boldsymbol{\Lambda}^{-1}(\mathbf{x} - \mathbf{x}')\right), \tag{3.20}$$

where $\sigma_m > 0$ is the signal variance, which determines how far the function $f(\mathbf{x})$ varies from the mean. $\boldsymbol{\Lambda}$ is a positive definite diagonal matrix of the length parameters, which influence how much the separation between $\mathbf{x}$ and $\mathbf{x}'$ will affect their covariance. Both σ_m and $\boldsymbol{\Lambda}$ are denoted as hyperparameters and can be optimized by maximizing the GP model's likelihood function [89].

To train a GP model with a set of input-output observations $(\boldsymbol{X}, \boldsymbol{Y})$, firstly the form of the kernel function is specified. Then its hyperparameters are optimized by maximizing the likelihood $p(\boldsymbol{Y}|\boldsymbol{X}, \sigma_m, \boldsymbol{\Lambda})$. At last the covariance matrix $\boldsymbol{K}(\boldsymbol{X}, \boldsymbol{X}')$ is determined using Eq. (3.20).

Based on the learned model with a prior zero-mean, given a set of new input $\mathbf{x}_*$, GP regression predicts the distribution of the function value $\boldsymbol{f}_*$, denoted as $p(\boldsymbol{f}_*|\boldsymbol{X}, \boldsymbol{Y}, \mathbf{x}_*)$ using the following equations:

$$\boldsymbol{\mu}_* = \boldsymbol{k}_*^T(\boldsymbol{K} + \sigma_n \boldsymbol{I})^{-1}\boldsymbol{Y}, \tag{3.21}$$

$$\boldsymbol{\Sigma}_* = \boldsymbol{k}_{**} - \boldsymbol{k}_*^T(\boldsymbol{K} + \sigma_n \boldsymbol{I})^{-1}\boldsymbol{k}_*, \tag{3.22}$$

where $\boldsymbol{\mu}_*, \boldsymbol{\Sigma}_*$ denote the expectation and variance of $\boldsymbol{f}_*$. $\boldsymbol{k}_*$ is the covariance matrix of $(\boldsymbol{X}, \mathbf{x}_*)$. $\boldsymbol{k}_{**}$ is the covariance matrix of $\mathbf{x}_*$ itself. σ_n represents the output noise.

An essential feature of GP regression is that it provides not only a prediction $\boldsymbol{\mu}_*$ but also its uncertainty $\boldsymbol{\Sigma}_*$. This information can be further used to design proper robot behaviours. For instance, if the prediction uncertainty is high, which can be caused by unnatural or new human motion types that not have been seen by the robot before. In this case, the robot should move more carefully, e.g. reduces its speed or control gain.

Note that the definition in Eq. (3.17) results in a GP model with multiple outputs which is more challenging to handle in comparison with the single-output case. A common way to simplify the problem is to assume that, given an input $\mathbf{x}$, each component of the N- dimensional output $y_1, ... y_N$ is independent. With this assumption, it is possible to treat each output separately, i.e. learning the functional relationship $y_i = \boldsymbol{f}_i(\mathbf{x}) + \sigma_i$ for each output [83].

The GP model with the definition of input and output above only allows a one-step prediction. In practice, sometimes it is desired to achieve a multi-step prediction, i.e. to determine $\hat{\mathbf{y}}(k+1), ... \hat{\mathbf{y}}(k+N)$ at time point t_k for N prediction steps so that more information of human motion trend can be gathered and later used for robot planning. However, according to Eq. (3.17), this is only possible if input signals $\mathbf{x}(k+1), ... \mathbf{x}(k+N)$ are available. To deal with this problem, the predicted output $\hat{\mathbf{y}}(k+1)$, together with t_{k+1} is taken as a "virtual" input $\hat{\mathbf{x}}(k+1)$ of the next time instant. This process is repeated until the desired number of predictions is achieved.

3.3.2 Sparse Gaussian process

A well-known disadvantage of the GP regression is its high computational complexity of $\mathcal{O}(M^3)$ with M training points. To deal with this issue, sparse GP regression [83] is applied which makes use of a so-called inducing points set with a much smaller size, denoted as $\boldsymbol{X}_u$. The inducing points are chosen from the original input set by comparing the normalized squared distance between the new coming sample and each existing point in $\boldsymbol{X}_{u,i}$, which is determined as follows:

$$\left(\boldsymbol{X}_+ - \boldsymbol{X}_{u,i}\right)^T \boldsymbol{W} \left(\boldsymbol{X}_+ - \boldsymbol{X}_{u,i}\right), \tag{3.23}$$

where $\boldsymbol{X}_+$ represents the new incoming input, $\boldsymbol{W}$ is a positive definite weighting matrix. Only if the distance to any exciting inducing points exceeds a threshold, then $\boldsymbol{X}_+$ is added to $\boldsymbol{X}_u$.

Based on the characteristic of Gaussian distribution and the assumption of zero-mean, the posterior distribution of the inducing points set $p\left(\boldsymbol{f}_u|\boldsymbol{X}, \boldsymbol{Y}, \boldsymbol{X}_u\right)$ is determined as:

$$\boldsymbol{\mu}_u = \boldsymbol{k}_u^T(\boldsymbol{K} + \sigma_n\boldsymbol{I})^{-1}\boldsymbol{Y}, \tag{3.24}$$

$$\Sigma_u = \boldsymbol{k}_{uu} - \boldsymbol{k}_u^T(\boldsymbol{K} + \sigma_n\boldsymbol{I})^{-1}\boldsymbol{k}_u, \tag{3.25}$$

where $\boldsymbol{\mu}_u$, $\boldsymbol{\Sigma}_u$ denote the expectation and variance of $\boldsymbol{f}_u$, $\boldsymbol{k}_u$ represents the covariance of $(\boldsymbol{X}, \boldsymbol{X}_u)$, $\boldsymbol{k}_{uu}$ is the covariance of $\boldsymbol{X}_u$. Combining Eq. (3.24) and Eq. (3.25) with Eq. (3.21) and Eq. (3.22), the prediction equations are reformulated as follows:

$$\boldsymbol{\mu}_* = \boldsymbol{k}_*^T(\boldsymbol{K} + \sigma_n \boldsymbol{I})^{-1}\boldsymbol{Y} = \boldsymbol{k}_{*,u}^T \boldsymbol{k}_{uu}^{-1} \boldsymbol{k}_u^T(\boldsymbol{K} + \sigma_n \boldsymbol{I})^{-1}\boldsymbol{Y} = \boldsymbol{k}_{*,u}^T \boldsymbol{k}_{uu}^{-1}\boldsymbol{\mu}_u \tag{3.26}$$

$$\boldsymbol{\Sigma}_* = \boldsymbol{k}_{**} - \boldsymbol{k}_*^T(\boldsymbol{K} + \sigma_n \boldsymbol{I})^{-1}\boldsymbol{k}_* = \boldsymbol{k}_{**} - \boldsymbol{k}_{*,u}^T \boldsymbol{k}_{uu}^{-1} \boldsymbol{k}_u^T(\boldsymbol{K} + \sigma_n \boldsymbol{I})^{-1}\boldsymbol{k}_u \boldsymbol{k}_{uu}^{-1}\boldsymbol{k}_{*,u}$$

$$= \boldsymbol{k}_{**} - \boldsymbol{k}_{*,u}^T \boldsymbol{k}_{uu}^{-1}\left(\boldsymbol{k}_{uu} - \boldsymbol{\Sigma}_u\right)\boldsymbol{k}_{*,u}, \tag{3.27}$$

in which only the distributions of inducing points set (prior: $\boldsymbol{k}_{uu}$, posterior: $\boldsymbol{\mu}_u, \boldsymbol{\Sigma}_u$) are used to compute $\boldsymbol{\mu}_*$ and $\boldsymbol{\Sigma}_*$.

3.3.3 Online Sparse Gaussian process

As mentioned at the beginning of this chapter, this thesis mainly focuses on online human motion prediction. The primary goal is to predict human motion state for a finite time-horizon based on a model and real-time measurements. On the other hand, it is crucial to study how to take advantages of the incoming measurements to correct the model. For instance, the recursive least square method enables an online identification of the model parameters. The Kalman Filter keeps updating the error covariance. Similarly, a GP model can also be online adjusted via Bayesian state estimation technique.

In this thesis, a method proposed in [83], named as Sparse Online Noisy Input GP (SONIG) was implemented. It allows both a sparse representation of the training data and online processing of the inducing points set. The online updating rule was derived based in the recursive GP proposed in [90]. The algorithm is briefly summarized as follows.

Assuming that the size of the inducing points set $\boldsymbol{X}_u$ remains unchanged and the GP model has zero-mean, the aim is to recursively update its distribution $\boldsymbol{\mu}_u(k), \boldsymbol{\Sigma}_u(k)$ with new incoming data denoted as $(\boldsymbol{X}(k), \boldsymbol{Y}(k))$ at time point k. The whole procedure consist of three steps:

1. Inference: the estimated distribution of $\boldsymbol{f}(k)$ is calculated based on $\boldsymbol{\mu}_u(k-1)$ and $\boldsymbol{\Sigma}_u(k-1)$ from the previous step.

$$\boldsymbol{L}(k) = \boldsymbol{k}(\boldsymbol{X}(k), \boldsymbol{X}_u(k-1))\boldsymbol{k}(\boldsymbol{X}_u(k-1), \boldsymbol{X}_u(k-1))^{-1}, \tag{3.28}$$

$$\hat{\boldsymbol{\mu}}(k|k-1) = \boldsymbol{L}(k)\boldsymbol{\mu}_u(k-1), \tag{3.29}$$

$$\hat{\boldsymbol{\Sigma}}(k|k-1) = \boldsymbol{k}(\boldsymbol{X}(k), \boldsymbol{X}(k)) - \boldsymbol{L}(k)\boldsymbol{k}(\boldsymbol{X}_u(k-1), \boldsymbol{X}(k))$$

$$+ \boldsymbol{L}(k)\boldsymbol{\Sigma}_u(k-1)\boldsymbol{L}(k)^T. \tag{3.30}$$

2. Correction: the algorithm corrects the distribution of $\boldsymbol{f}(k)$ with the measurement $\boldsymbol{Y}(k)$. The principle is the same as for Kalman filter (see Eq. (2.8)).

$$\mathbf{K}(k) = \hat{\boldsymbol{\Sigma}}(k|k-1)\left(\hat{\boldsymbol{\Sigma}}(k|k-1) + \boldsymbol{\sigma}\right)^{-1}, \tag{3.31}$$

$$\boldsymbol{\mu}(k) = \hat{\boldsymbol{\mu}}(k|k-1) + \mathbf{K}(k)\left(\boldsymbol{Y}(k) - \hat{\boldsymbol{\mu}}(k|k-1)\right), \tag{3.32}$$

$$\boldsymbol{\Sigma}(k) = \hat{\boldsymbol{\Sigma}}(k|k-1) - \mathbf{K}(k)\hat{\boldsymbol{\Sigma}}(k|k-1). \tag{3.33}$$

3. Smoothing update: the corrected distribution $\boldsymbol{\mu}(k), \boldsymbol{\Sigma}(k)$ is passed backwards to update $\boldsymbol{\mu}_u(k), \boldsymbol{\Sigma}_u(k)$. The principle is similar to Kalman smoothing algorithm [70].

$$\tilde{\mathbf{K}}(k) = \boldsymbol{\Sigma}_u(k-1)\boldsymbol{L}^T(k)\left(\hat{\boldsymbol{\Sigma}}(k|k-1) + \boldsymbol{\sigma}\right)^{-1} \tag{3.34}$$

$$\boldsymbol{\mu}_u(k) = \boldsymbol{\mu}_u(k-1) + \tilde{\mathbf{K}}(k)\left(\boldsymbol{Y}(k) - \boldsymbol{\mu}(k|k-1)\right), \tag{3.35}$$

$$\boldsymbol{\Sigma}_u(k) = \boldsymbol{\Sigma}_u(k-1) - \tilde{\mathbf{K}}(k)\boldsymbol{L}(k)\boldsymbol{\Sigma}_u(k-1). \tag{3.36}$$

Note that even not considered in this thesis, this algorithm can be extended for online learning of the hyperparameters [90].

3.3.4 Evaluation and discussion

Fig. 3.3 presents the workflow of human motion prediction based on online sparse Gaussian Process Regression. The procedure is divided into two phases. The offline training phase aims to obtain the optimal hyperparameters and generate an initial distribution of the inducing points set. In standard GP, the input is assumed to be noise-free. However, it does not hold in this work since the input contains position and velocity measurements. Hence, the GP training method proposed in [91] with consideration of (Gaussian) input noise is implemented. In the next step, the inducing points are selected by comparing the relative distance defined in Eq. (3.23). Lastly, an initial distribution of the inducing point set is determined using Eq. (3.24) and Eq. (3.25).

The online phase consists of a multi-step motion prediction and a recursive update for the inducing points set. When an observation $\mathbf{x}(k)$ is available, the algorithm firstly performs a one-step prediction $\hat{\mathbf{x}}(k)$ using Eq. (3.21) and (3.22), then takes the predicted value as a virtual input for the next iteration until the number of prediction steps N is reached. Lastly, the update algorithm recomputes the distribution of the inducing points.

The recorded motion data in the table-top manipulation task (Fig. 2.1(a))) is used for evaluation. Since the raw data only contains position measurements, velocity and

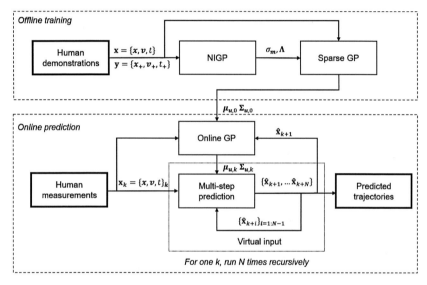

Figure 3.3: Flowchart of human motion prediction based on online sparse Gaussian Process Regression

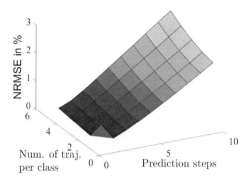

Figure 3.4: Normalized root mean square error of the motion prediction with online sparse GP regression over number of trajectories in each motion class and prediction steps.

acceleration are determined based on the constant acceleration model with KF. The data is randomly separated into a training and a test set. The motion trajectories are classified by their initial and target positions. Totally 9 motion classes are considered.

Impact of training set size and prediction horizon

Firstly the dependency of prediction error on training set size and prediction horizon was studied. The evaluation was performed by varying two parameters: 1) the number of trajectories in each motion class for training (ranging from 1 to 5), 2) the number of prediction horizons (ranging from 1 to 10). Performance was quantified by NRMSE and plotted in Fig. 3.4.

The results show no significant difference in prediction error when increasing the number of trajectories for training in each motion class from 1 to 5. They suggest that the algorithm does not require a large number of demonstrations to learn a GP model that is efficient enough for prediction. However, authors in [83] drew opposite conclusions with the same prediction algorithm. They found that more training points provide higher accuracy. Nevertheless, they mentioned that the results strongly rely on the data set and the system function itself. Hence, no general conclusion on the impact of the training set size can be made based on the study in this thesis.

Moreover, the prediction error increases almost linearly to the prediction step and re-

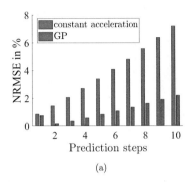
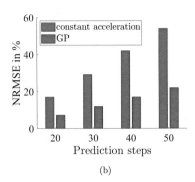

(a)   (b)

Figure 3.5: Performance of the short- and long-term online prediction based on constant
           acceleration (blue) and Gaussian Process (red) in a pick-and-place task.

mains at a low level ($< 3\%$ for 10 prediction steps).

The advantage of Sparse GP was confirmed in the simulation. During the implementation, after performing 1305 training points, the size of the inducing points set was still kept small (18), making online computation more efficient than standard GP regression.

Performance in short- and long-term prediction

Similar to the previous chapter, performance in both short- and long-term human motion prediction is evaluated. Fig. 3.5 shows the results in comparison to the constant acceleration model with KF. In both cases the prediction error based on online GP regression is much lower. Details of the predicted trajectories in one single demonstration are shown in Fig. 3.6.

The first factor that contributes to the result is GP's ability to fit non-linear functions. Due to its non-parametric characteristic, GP extends the Bayesian linear regression to approximate a more flexible class of models [89]. Therefore, it is especially beneficial for long-term prediction, in which a sufficiently exact approximation of the motion pattern is required. Another contributing factor is the data set. Since the motion trajectories in training and test sets are similar, there exist no significant uncertainties. Moreover, the velocity and acceleration values are pre-filtered through KF, which suppressed the influence of noise.

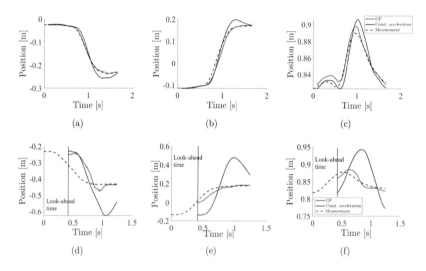

Figure 3.6: Prediction in one single demonstration based in constant acceleration (blue) and Gaussian process (red) with $N = 10$ (first row) and $N = 50$ (second row):(a),(d) $x-$dimension, (b),(e) $y-$dimension, (c),(f) $z-$dimension.

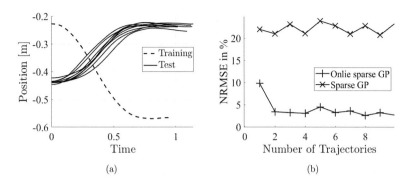

Figure 3.7: Simulation with new test set: (a) comparison between the training (dashed) and test trajectories (solid), (b) prediction error over the number of observed trajectories.

Impact of online GP

As introduced before, one advantage of online Sparse GP regression is that it can recursively incorporate new measurements as training points to refine the inducing points distribution. In order to validate this statement, another simulation is performed using a test set containing motion types that are not included in the training set.

As plotted in Fig. 3.7(a), trajectories in the test set are with new initial and target positions, and the movement direction is opposite to the demonstrated trajectory. Fig. 3.7(b) shows that the prediction with both standard and online Sparse GP did not work well in the first trial since the correlation between the new observations and learned initial distributions was small. Nevertheless, the corrective effect of online GP is observable in Fig. 3.8(d). The moment in which the predicted value abruptly changed is when a new point is added and the distribution of the inducing point set is recomputed. After processing several measurements, the performance is getting better, and the uncertainty becomes smaller, see Fig. 3.8. In comparison, there is no improvement by standard Sparse GP.

This simulation shows a promising feature of online GP in human motion prediction, namely fast adaptability to unknown motion classes. Accordingly, the size of training set can also be reduced. Nevertheless, the algorithm still requires several demonstrations to learn a new motion type, and the prediction error in the first few iterations is large.

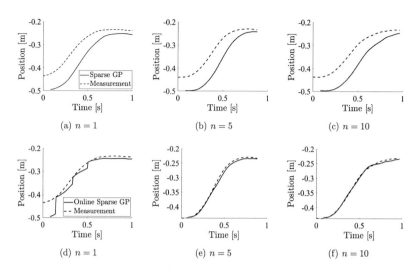

Figure 3.8: Prediction in $x-$ dimension with horizon $N = 10$ after performing n test trajectories: (a)-(c) Sparse GP, (d)-(f) Online Sparse GP.

3.3.5 Gaussian process with explicit basis functions

So far, all the derivations are performed under the assumption of a zero-mean function. It reduces the computational complexity and usually works well if predictions are based on inputs that are similar to the ones in the training set [89]. However, it is not always sufficient to assume a zero mean. As shown in the last section, just because the initial and target positions are changed, GP needs to retrain its model by observing several demonstrations even for the same motion type. It also happens if the human performs the same task with a different duration. The main reason is that a zero-mean can only adapt to such variations by updating its posterior distribution, which is less effective.

This disadvantage can be partly compensated if some prior knowledge of the system is available. For instance, relations between variables such as position, velocity or force must satisfy basic physical laws. Another example is that in the context of human-robot collaboration, humans and robots are normally coupled through their kinematic and dynamic constraints. Incorporating such effects in some explicit basis functions can increase the interpretability and generalizability of the GP model.

A GP model with explicit basis functions is formalized as follows:

$$\mathbf{y} = \boldsymbol{h}(\mathbf{x})^T\boldsymbol{\beta} + \mathbf{g}(\mathbf{x}) + \boldsymbol{\sigma}, \quad \mathbf{g}(\mathbf{x}) \sim \mathcal{GP}(0, \boldsymbol{k}\,(\mathbf{x}, \mathbf{x}')), \ \boldsymbol{\sigma} \sim \mathcal{N}(0, \sigma_n^2 \boldsymbol{I}), \qquad (3.37)$$

where $\mathbf{g}(\mathbf{x})$ is a zero-mean GP, $\boldsymbol{h}(\mathbf{x})$ is a set of basis functions, $\boldsymbol{\beta}$ represents the coefficients of the basis functions, which can be either pre-fixed or optimized together with the hyperparameters in $\boldsymbol{k}$.

Note that this formulation is similar to a linear regression problem, the main difference is that the residual, namely $\mathbf{g}(\mathbf{x})$ is modeled as a GP.

Denoting $\boldsymbol{f}\,(\mathbf{x}) = \boldsymbol{h}(\mathbf{x})^T\boldsymbol{\beta} + \mathbf{g}(\mathbf{x})$, the prediction of $\boldsymbol{f}\,(\mathbf{x}_*)$ with new input $\mathbf{x}_*$ based on a training set $(\boldsymbol{X}, \boldsymbol{Y})$ is then calculated by:

$$\boldsymbol{\mu}_* = \boldsymbol{h}(\mathbf{x}_*)^T\boldsymbol{\beta} + \boldsymbol{k}_*^T(\boldsymbol{k} + \sigma_n \boldsymbol{I})^{-1}(\boldsymbol{Y} - \boldsymbol{H}^T\boldsymbol{\beta}), \qquad (3.38)$$

$$\boldsymbol{\Sigma}_* = \boldsymbol{k}_{**} - \boldsymbol{k}_*^T(\boldsymbol{k} + \sigma_n \boldsymbol{I})^{-1}\boldsymbol{k}_*. \qquad (3.39)$$

For simplification, the parameter $\boldsymbol{\beta}$ is assumed to be deterministic. Nevertheless, it is possible to consider its uncertainty by regarding it as a Gaussian distribution, i.e. $\boldsymbol{\beta} \sim \mathcal{N}(\boldsymbol{b}, \boldsymbol{B})$, then the GP model turns to:

$$\boldsymbol{f}(\mathbf{x}) \sim \mathcal{GP}\left(\boldsymbol{h}(\mathbf{x})^T\boldsymbol{b}, \boldsymbol{k}\,(\mathbf{x}, \mathbf{x}') + \boldsymbol{h}(\mathbf{x})^T\boldsymbol{B}\boldsymbol{h}(\mathbf{x})\right), \qquad (3.40)$$

In comparison to the standard form, the covariance function includes an extra non-negative term caused by the uncertainty in parameter $\boldsymbol{\beta}$. Both its mean $\boldsymbol{b}$ and covariance $\boldsymbol{B}$ can be learned together with the hyperparameters by maximizing the likelihood function. Accordingly, the predictive equation is also extended with the covariance function. More details can be found in [89]. It should be pointed out that a GP with explicit basis functions is not a purely data-driven approach any more, since physical knowledge has been included in the model. The concept of hybrid physical and data driven approach will be discussed in detail in the next chapter.

A simulation study is performed to investigate whether incorporating an explicit basis function will enhance the prediction performance in case of varying initial and target positions. Two online sparse GPs (one zero-mean and the other with basis function) are trained to predict human reaching motion with new initial and target positions. According to the previous input-output definition, a constant velocity model is chosen as the basis function, in which the human hand velocity is assumed constant within one sampling period. It yields:

$$\underbrace{\begin{pmatrix} \boldsymbol{x}\,(k+1) \\ \boldsymbol{v}\,(k+1) \end{pmatrix}}_{\boldsymbol{y}} = \underbrace{\begin{pmatrix} \boldsymbol{I} & T_s\boldsymbol{I} & 0 \\ 0 & \boldsymbol{I} & 0 \end{pmatrix}}_{\boldsymbol{\beta}^T} \underbrace{\begin{pmatrix} \boldsymbol{x}\,(k) \\ \boldsymbol{v}\,(k) \\ t_k \end{pmatrix}}_{\boldsymbol{h}(\mathbf{x})}. \qquad (3.41)$$

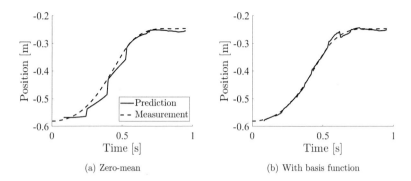

Figure 3.9: Prediction with online- (a) comparison between the training (dashed) and test trajectories (solid), (b) prediction error over the number of observed trajectories.

Fig. 3.9 illustrates the benefit of incorporating a physical-based basis function in dealing with new motion trajectories that are not included in the training set. It can be seen that the ordinary Sparse GP takes a few iterations to adapt to the new incoming data. During this procedure, the prediction "jumps" as the distribution of inducing set updates. In caparison, the physical-based basis function utilizes the prior knowledge of the system and reacts fast to the motion variation. So GP only needs to handle the residuals that the basis function cannot cover. This concept improves the model's flexibility and makes the predicted trajectory more smooth .

Nevertheless, it should be noted that incorporating improper basis functions in a GP can bring additional error sources and uncertainties. Another simulation was performed with a test set that contains similar trajectories as in training set to complete the analysis. As shown before, in this case, a zero-mean GP has already provided adequate performance. Fig. 3.10 shows that the GP with basis function does not consistently outperform the zero-mean GP, especially for long-term predictions. It is not surprising since the basis function, namely the constant velocity model, can only describe the short-term transition behaviour of the motion variables. Hence, it introduces more significant errors in long-term prediction, and the overall performance is reduced.

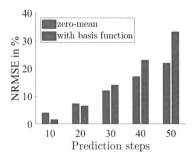

Figure 3.10: Comparison of prediction error between zero-mean GP (blue) and GP with basis function (red).

3.4 Summary

In this chapter, two data-driven approaches, namely inverse optimal control (IOC) and Gaussian Process (GP) regression, were studied and evaluated by prediction of human reaching motion. The aim is to utilize the benefits of machine learning techniques in modelling complex or nonlinear systems under uncertainties.

The study on IOC is motivated by the assumption that the human central nervous system synthesizes motion by minimizing a cost function over time. The cost function defined in this work is quadratic and consists of three features: distance to the target position, end-effector velocity and acceleration. The purpose is to identify their weightings based on observed human trajectories. The IOC problem is formulated based on the maximum entropy principle and solved by a numerical solver for nonlinear optimization. Afterwards, prediction of human motion is performed by online solving the optimization problem with the identified cost function.

Comparing to simple kinematic models introduced in the previous chapter, prediction based on IOC has an edge in long-term predictions. Moreover, results show that a large data set is not required for learning a reaching movement. On the other side, some limitations/problems of the IOC method have been detected during the implementation. The most important ones are: (1) The parameterization of the cost function in a linear quadratic form with only three features may not be enough to describe human motion generation synthesis. (2) There exist large uncertainties in the identified parameters. The uniqueness of the learned cost function cannot be guaranteed. (3) Too few features have been included in the cost function, especially the arm joint movements.

For future work, besides improving the framework in consideration of the above listed problems, it is worth investigating online inverse optimal control, in which the objective function is learned sequentially from pairs of input and output observations. Combining IOC with model-based optimal control methods (e.g. Model Predictive Control) to develop a learning-based control framework can also be an promising research topic.

The motivation to study GP regression is its non-parametric characteristic, i.e. not limited by a fixed functional form. A well-known disadvantage of the standard GP is its computational complexity, which is critical for online use. In order to overcome this issue and achieve a real-time adaptation to incoming measurements, an online sparse GP regression algorithm was applied. Results show that GP outperforms the other methods implemented so far in both short-term and long-term predictions. Similar to the IOC method, a small-size training set is sufficient to achieve satisfied prediction quality. The sparse approach further reduces the computational time, especially for a multi-step prediction. Moreover, the online sparse GP has a clear advantage in dealing with new data from unknown batches that are not included in the training set.

Like many other purely data-driven approaches, a prior zero-mean GP suffers problems when making predictions with data that not has been seen yet. Despite its adaptability, performance in the first iterations is usually not good. A possible solution is to incorporate several basis functions, particularly physical models, to provide more prior knowledge of the system. Some preliminary results in this work show that adding a physics-based basis function helps enhance the prediction performance, but only in the region where the physical model represents well. Improper choice of basis functions may bring additional error biases in the system.

One limitation of the study is that all the analyses were performed using a data set that only includes one motion type. Even though the trajectories are recorded from different participants and have varying initial and final positions, they all belong to unconstrained point-to-point reaching motion. Moreover, only the translational motion has been considered. Future work should consider various motion types and more degrees of freedom.

A combination of physical and data-driven modelling approaches to utilize their strength is becoming a new research trend. This topic will be discussed in the next chapter.

4 Human Motion Prediction based on Dynamic Movement Primitives

4.1 Introduction

4.1.1 Hybrid physical and data-driven approach

In the last two chapters, the pros and cons of the kinematic and data-driven models are discussed. Simple kinematic models, such as constant velocity or constant acceleration models, have a clear physical interpretation, provide satisfying performance in short-term predictions, and are easy to implement. However, they usually do not work well for long-term predictions since the key assumption, namely a piecewise constant state with white noise, is only valid in a small time interval. Complex kinematic models, e.g. biomechanical models, provide better performance in human motion reconstruction but are computationally expensive and the parameter identification is difficult. In comparison, data-driven models aim to learn a movement pattern from human demonstrated trajectories using different function approximators, e.g. neural networks, Hidden Markov Models, GP, etc. Many of these methods are derived from Bayes'theorem and can therefore deal with uncertainties. Nevertheless, the model accuracy depends on the quality of data (e.g., noise level), the type of function approximators, and the learning algorithm. The offline complexity is high, including training data pre-processing (trajectory segmentation, signal filtering, etc.), choice of function approximators, design of learning algorithms, and parameter tuning. Moreover, one fundamental question still remains in data-driven approaches: how large should the training set be to achieve a generalizable model?

In recent years, hybrid physical and data-driven model has become a new research trend in human modeling, which attempts to utilize the strength of both approaches. For instance, authors in [46] propose a latent force model, in which the system is represented by a mechanic model that driven by a low dimensional "latent force" (Fig. 4.1(a)). Each latent force is taken to be independent and descried by GP. Several follow-up publications extend the model in different ways, including non-linear latent force model [92],

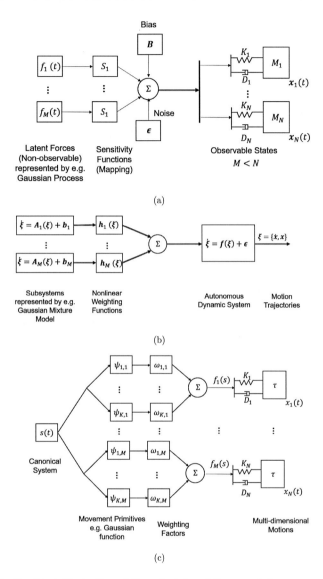

Figure 4.1: Graphical illustration of the principles of three modeling approaches: (a) latent force model, (b) dynamic system based model, (c) dynamic movement primitives.

stoachsic latent force model [93] and switching dynamic latent force model [94]. Another widely studied approach is the dynamic system (DS)- based method, which was first proposed in [47] and comprehensively discussed in [95]. The key idea is to model a motion as an autonomous nonlinear time-invariant first-order dynamical system (Fig. 4.1(b)). Its transition function is represented by a sum of subsystems with nonlinear weighting functions, and learned by a probabilistic framework. This method has been implemented in various human-robot interaction tasks to generate adaptive robot motion, especially in physical human-robot-collaboration [96, 51].

Dynamic Movement Primitives (DMP), motivated by motor control of biological systems, has been drawing more and more attention in robotic fields [97]. Initially developed for imitation learning, DMP extends a physically well-understood linear attractor dynamic with a learnable non-linear forcing term to flexibly adapt to different movement profiles. The forcing term consists of a set of basis functions driven by a convergent canonical system (Fig. 4.1(c)). Besides the conventional DMP formulation, various extensions have been made. Representative examples are Probabilistic Movement Primitives (ProMPs) [98] to handle uncertainties, DMP with potential field modulation [99] for online obstacle avoidance, and adaptive DMP [100, 101] to deal with the environment and task variation. The main difference between GS and DMP is that DMP learns a model for each dimension separately0, while DS learns a multi-dimensional model [95].

Fig. 4.1 briefly summarizes the principles of the three hybrid modelling approaches introduced above. It can be seen that their fundamental ideas are similar: using an (overly simplified) mechanical model (usually described by first or second differential equations) to roughly cover the transitions of the observable physical states (e.g. position, velocity), then refining the model through machine learning techniques to achieve a sufficiently exact approximation of the system's dynamic. Within this concept, the mechanical term can make predictions in the region where no training data is available. Furthermore, the data-driven term enables the description of system dynamics, which are too complex for a purely mechanical modelling approach.

In this chapter, DMP-based modeling and prediction of human motion are studied. As discussed in the previous chapter, although online GP has already provided satisfactory performance, it still has problems in generalization, especially when making predictions with unseen data. In comparison, DMP has several advantages, including:

- a simple dynamic system to interpret goal-directed behavior with the guarantee of stability,

- temporal and spatial scalability,

- real-time computation as well as arbitrary modulation of trajectories,

- ability to incorporate coupling terms, e.g., for motion synchronization.

Despite these advantages, conventional DMP still has some limitations, in particular:

- All the parameters of the basis functions to construct the shape-attractor need to be manually adjusted, and the assumption of a linear combination of these basis functions may not always hold.

- The shape-attractor term vanishes too fast due to the exponential convergence of the canonical system.

- The conventional spatial scaling method may fail if the start and target positions are close to each other.

In order to overcome these limitations, a novel DMP formulation for learning human trajectories is proposed in this chapter with the following extensions:

- a Gaussian Process (GP) based shape-attraction term,

- a state-dependent weighting function to control the transition of goal-attraction and shape-attraction terms,

- a rotation-based representation of the spatial scaling.

The proposed formulation is validated based on the recorded motion data set and compared with the GP regression method introduced in the previous chapter.

4.1.2 Modeling of interaction dynamics

So far, various modeling and prediction approaches are studied in the context of an individual agent. Note that collaboration is a joint activity in which all participants are coupled by physical or task constraints and keep interacting with each other. Hence, it is crucial to understand and interpret the coupling effects and the interactive dynamics. Taking handover as an example, it means correlations between the motions of the giver and receiver.

Probabilistic approaches provide a perspective solution by correlating motion trajectories with their joint distributions. The authors in [102] presented a new representation named Interaction Primitives, which is built on the framework of DMP. The key idea is to correlate the two agents by a probabilistic joint distribution of their DMP parameters. Later in [103], the authors integrated Interaction Primitives into the structure of Probabilistic Movement Primitives (ProMPs), which enables learning correlations between the trajectories of different agents in one probabilistic framework. Both works were validated in human-robot handover experiments. The result show that the Interactive Primitives improve the robustness of the robot motion against task variation, especially when human changed the handover location.

The same question still remains in such probabilistic approach how large should the data set be to achieve a generalizable representation of the interactive behavior? The authors in [104] proposed an active learning framework, which computes the most crucial task instance that human should demonstrate to learn a library of ProMPs. Their results show a better generalization with fewer demonstrations than other random sampling methods. Nevertheless, the framework only works in a fixed region. It is suitable for tabletop manipulation tasks but still has difficulties achieving a flexible task that is not limited in space.

In this thesis, a extended DMP formulation is presented to describe the interactive dynamics. The model is developed for point-to-point reaching motion, but can be extended to other collaboration tasks which aim to maintain or modify a task-depending formation. A similar concept is proposed in [105], where the authors interpret the interaction dynamic between two agents as two coupled oscillators via an impedance in the phase plane. A one-dimensional phase variable is used to describe the task execution progress. Correlations between the phase variable and the end-effector kinematics are represented by probabilistic distributions. Unlike their concept, in this work it is assumed that the agents are coupled by a virtual force. Their motions are generated by the sum of a primary force in their own DMPs (the shape-attraction term) and a state- dependent coupling force (the goal attraction term). Similar interpretations can be found in [106] for describing the interaction in multi-robot formation control, and in [107] for physical human-robot collaboration. In this thesis, to deal with human uncertainties, a probabilistic model (GP) is used to learn the shape-attraction term.

The remainder of the chapter is organized as follows. The regular form of DMP and its temporal and spatial invariance property is briefly reviewed in Section 4.2. Section 4.3 describes the details of the proposed DMP formulation. Section 4.4 presents the learning procedure of a DMP model based on human demonstrations. Evaluation of the approach with measurement data is shown in Section 4.5. Section 4.6 describes an interaction model based on the DMP formulation. Finally, Section 4.7 provides a summary of this chapter.

4.2 Regular DMP

DMP is originally developed to model attractor behaviors of nonlinear dynamical systems [48]. Its mathematical formulation is presented as follows:

$$\tau \dot{v} = K\left(g - x\right) - Dv + \left(g - x_0\right)f(s), \tag{4.1}$$

where τ is a temporal scaling factor. x, v represent position and velocity. g, x_0 represent target and initial position. $K, D \succ 0$ are gain matrices. f is a nonlinear forcing term. $s \in [0, 1]$ is defined as a phase variable, which determines the accomplishment of the motion trajectory. Typically, one has $s = 1$ for $x = x_0$ and $s = 0$ for $x = g$. Its dynamic is described by the following canonical system:

$$\tau \dot{s} = -\alpha s, \tag{4.2}$$

where $\alpha > 0$ determines the rate of convergence of s.

Eq. (4.1) shows that the DMP formalism consists of two terms: a linear PD-type point attractor and a non-linear forcing function, or denoted as shape attractor. The former guarantees convergence of position x to g. The latter determines the shape of the motion trajectory. One essential characteristic of f is that its value does not rely on time and motion parameters. Hence, it can be scaled temporally by τ and spatially by $(g - x_0)$, without losing the basic form of the motion trajectory. Authors in [48] named it as the invariance property of DMP. With this property, one can use one single DMP model to generate several trajectories with similar forms by simply modifying the temporal and spatial scaling parameters. A graphical illustration of the invariance property is shown in Fig. 4.2.

The shape attractor f decides the motion profile and can be chosen either phasic (for point-to-point motion) or periodic (for oscillate motion). In principle, f can be described by an arbitrary non-linear approximator. However, it must satisfy $f(s \to 0) \to 0$ so that the whole system converges and satisfies $x \to g$. In order to cover more versatile dynamics, f is usually formulated as a weighted linear combination of various basis functions, such as:

$$f_i(s) = \frac{\sum_{j=1}^{N} \psi_j(s)\Omega_j}{\sum_{j=1}^{N} \psi_j(s)}s, \tag{4.3}$$

where i stands for $i-$th dimension of f, ψ are fixed basis functions and Ω are adjustable weights, N is the total number of basis functions.

The Gaussian function is one of the most commonly used candidates as a basis function ψ. It has a squared exponential form:

$$\psi_j(s) = \exp\left(-\frac{1}{2b_j^2}(s - c_j)^2\right), \tag{4.4}$$

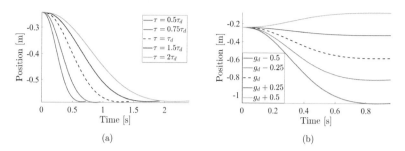

Figure 4.2: Graphical illustration of the invariance property of DMP, the dashed line represents the demonstrated trajectory with temporal scaling factor τ_d and goal position g_d: (a) generalization with varying τ , (b) generalization with varying g.

where c, b are the mean and variance of the Gaussian function.

4.3 Modified DMP

4.3.1 DMP with Gaussian process

Learning the shape attractor of a DMP model determines each basis function's weight Ω_j. The most commonly used method is the locally weighted regression (LWR). The advantages include fast learning procedure and stable parameterization [48]. One issue is that the parameters of each basis function, namely the mean c_j and variance b_j, need to be manually adjusted. As shown in the previous chapter, GP regression can approximate non-linear functions and has the advantage that all its hyperparameters are optimized automatically by maximizing the likelihood. Thus, in this work, each dimension of the non-linear forcing term $\boldsymbol{f} \in \mathbb{R}^3$ is formulated by GP with the phase variable s as input:

$$f_i \sim \mathcal{GP}\left(m_i\left(s\right), k_i\left(s, s'\right)\right), \quad i = 1, 2, 3 \tag{4.5}$$

with a squared exponential kernel function described in Eq. (3.20).

The sparse and online GP introduced in the previous chapter can also be implemented when learning a DMP model. The former aims to reduce the online computational load, the latter aims to adapt the GP model with new measurements.

One disadvantage of the above GP formulation is that the condition $(s \to 0 \Rightarrow \boldsymbol{f} \to \boldsymbol{0})$ cannot be guaranteed. To deal with this problem, the authors in [108] used $\ln(s)$ as input so that the squared exponential kernel function becomes zero for $\ln(s) \to -\infty$. Later in this work, an weighting function is added in front of $\boldsymbol{f}$ (see Eq. (4.14)). The zero-convergence of $\boldsymbol{f}$ is then achieved by forcing the weighting factor to converge to zero.

Relationship to GP with basis functions

Substituting Eq. (4.5) into Eq. 4.1 results in a combined DMP-GP model with its non-linear forcing term described by GP. In comparison to the GP with explicit basis function in Eq. (3.37), both formulations have some similarities. They both consist of a function to express prior information of the system, and a zero-mean GP to model the residuals. The DMP-GP model can be regarded as a special form of Eq. (3.37), in which the basis functions are represented by a second-order mass-spring-damper model, and all the dimensions are decoupled, i.e. all matrices are diagonal. The main difference is that in the DMP-GP model, the basis function and GP residuals are defined with different inputs. The former takes position and velocity as input, while the latter uses the phase variable.

4.3.2 Rotation based spatial scaling

In the standard form of DMP (Eq. (4.1)), the spatial scaling of the shape attractor $\boldsymbol{f}$ takes place through $\mathrm{diag}\,(\boldsymbol{g} - \boldsymbol{x_0})$, which means the scaling in each dimension is independent and scalar. A well-known problem is that if $g_i - x_{0,i}$ in dimension i is near to zero, then the non-linear term has almost no contribution due to the scaling. The bio-inspired DMP proposed in [99] has been developed to fixes that problem. The mathematical formulation is shown as follows:

$$\tau \dot{\boldsymbol{v}} = \boldsymbol{K}\,(\boldsymbol{g} - \boldsymbol{x}) - \boldsymbol{D}\boldsymbol{v} - \boldsymbol{K}\,(\boldsymbol{g} - \boldsymbol{x_0})\,s + \boldsymbol{K}\boldsymbol{f}(s). \tag{4.6}$$

In comparison to the standard DMP, the spatial scaling of the non-linear shape attractor is now achieved by adding (instead of multiplying) an additional term $(\boldsymbol{g} - \boldsymbol{x_0})\,s$. This design is beneficial in dealing with small values of $\boldsymbol{g} - \boldsymbol{x_0}$. However, the global scaling property is lost. It may cause problems in generating new motion trajectories, especially when the moving direction varies a lot.

To deal with this issue, authors in [109] comprehensibly analyzed DMP's spatial scaling properties and proposed a novel formulation based on a rotation matrix of two vectors.

Assuming that a nominal trajectory described by DMP has an initial position $\boldsymbol{x}_{0,n}$, a target position $\boldsymbol{g}_n$ and a forcing term $\boldsymbol{f}_n$. To generate a new motion trajectory with another initial position $\boldsymbol{x}_0$ and target position $\boldsymbol{g}$, the forcing term is scaled as follows:

$$\boldsymbol{f}(s) = \gamma \boldsymbol{R} \boldsymbol{f}_n(s), \tag{4.7}$$

where

$$\gamma = \frac{\|\boldsymbol{g} - \boldsymbol{x}_0\|}{\|\boldsymbol{g}_n - \boldsymbol{x}_{0,n}\|} \tag{4.8}$$

is the amplitude scaling factor. $\boldsymbol{R} \in \mathbb{R}^{3 \times 3}$ denotes the rotation matrix from vector $\boldsymbol{n}_0$ to $\boldsymbol{n}$, where

$$\boldsymbol{n}_0 = \frac{\boldsymbol{g}_n - \boldsymbol{x}_{0,n}}{\|\boldsymbol{g}_n - \boldsymbol{x}_{0,n}\|}, \boldsymbol{n} = \frac{\boldsymbol{g} - \boldsymbol{x}_0}{\|\boldsymbol{g} - \boldsymbol{x}_0\|}. \tag{4.9}$$

Denoting $\theta = \langle \boldsymbol{n}_0, \boldsymbol{n} \rangle$, the rotation matrix $\boldsymbol{R}$ is calculated based on Rodrigues' rotation formula:

$$\boldsymbol{R} = \begin{cases} -\boldsymbol{I}, & \text{if } \theta = \pi \\ \boldsymbol{I} + (\sin\theta)\,\boldsymbol{S}\,(\boldsymbol{\kappa}) + (1 - \cos\theta)\,\boldsymbol{S}^2\,(\boldsymbol{\kappa}), & \text{else .} \end{cases} \tag{4.10}$$

where $\boldsymbol{I} \in \mathbb{R}^{3 \times 3}$ is the identity matrix, $\boldsymbol{S}$ is a skew-symmetric matrix ,$\boldsymbol{\kappa}$ represents the unit vector of $\boldsymbol{n}_0 \times \boldsymbol{n}$.

The key idea is to formulate the spatial scaling as a three-dimensional rotation. It solves the problem when the initial and target positions are close in some directions without losing the general scaling property. The only case in which this method cannot work is when both initial and target positions are identical in all dimensions.

4.3.3 Weighting factor

An interesting and still open question is how to weight the contribution of linear attractor and the non-linear modulation in DMP formalism. In the standard DMP formulation, they are weighted by the phase variable s, since the non-linear term is usually multiplied by s. As discussed in [110], one disadvantage is that the non-linear term decreases too fast as s converges exponentially to zero. This makes the motion generated by DMP sensitive to the linear attractor's variation, especially the change of goal position.

Therefore, authors in [110] proposed a weighting function decoupled from the phase variable s. The function is formulated based on a Gauss error function with time t as variable. The key idea is that the non-linear term should be weighted more at the start of the motion (small t), and the linear feedback term should be weighted more at the end of the motion (large t). The transition of its value from 0 to 1 is determined by the mean and standard deviation of the Gaussian distribution. The former decides when the transition will occur. The latter is for the duration of the transition.

However, from a practical point of view, the time-dependent weighting function has following limitations:

- The shift of the weight depends on the total task execution time, which needs to be fixed a priori. However, it can be problematic in practice and limit flexibility.

- Since the time variable increases monotonically, the shift of the weights is not invertible, i.e. the non-linear forcing term's weight keeps reducing.

In considering these issues, in this thesis, a position-dependent weighting function is defined as follows:

$$\omega\left(\tilde{\boldsymbol{x}}\right) = 0.5 \left(1 + \mathrm{erf}\left(\frac{\|\tilde{\boldsymbol{x}}\| - \mu}{\sqrt{2}\epsilon}\right)\right), \qquad (4.11)$$

where erf stands for Gauss error function:

$$\mathrm{erf}(x) = \frac{2}{\sqrt{\pi}} \int_0^x \exp^{-t^2} \mathrm{d}t, \qquad (4.12)$$

and

$$\tilde{\boldsymbol{x}} = \boldsymbol{g} - \boldsymbol{x}. \qquad (4.13)$$

Note that Eq. (4.11) has the same form as in [110], except that the distance $\|\tilde{\boldsymbol{x}}\|$ between actual and target position is used as variable in Gauss error function instead of time t. Accordingly the mean value μ is also a distance parameter. The reason is that it makes more sense to judge the accomplishment of a reaching movement by observing the position instead of counting the time. Additionally, it allows the shift of the weights in two directions. Fig. 4.3 shows several variations of $\omega\left(\tilde{\boldsymbol{x}}\right)$ by changing these two parameters.

Based on the weighting function and the scaling method in Eq. (4.7), the DMP model proposed in this thesis is formulated as follows:

$$\tau\dot{\boldsymbol{v}} = \left(1 - \omega\left(\tilde{\boldsymbol{x}}\right)\right)\left(\boldsymbol{K}\tilde{\boldsymbol{x}} + \boldsymbol{D}\left(\boldsymbol{v}_g - \boldsymbol{v}\right)\right) + \omega\left(\tilde{\boldsymbol{x}}\right)\gamma\boldsymbol{R}\boldsymbol{f}(s). \qquad (4.14)$$

Note that to deal with a moving goal, a velocity feedback $\boldsymbol{v}_g$ of the target is added in the linear goal attraction term.

This equation shows that at the beginning of handover, since the distance between human and robot is large, $\omega\left(\tilde{\boldsymbol{x}}\right) \rightarrow 1$, i.e. the shape- attraction term dominates the motion. Since the robot has no prior information on the handover location, it plans the motion based on the learned trajectories from human demonstrations. As the distance becomes small, ω goes to zero so that the linear feedback goal-attraction term takes over and aims to track the movement of the human hand.

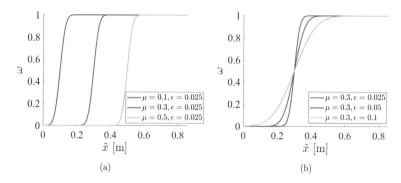

(a)   (b)

Figure 4.3: Graphical illustration of the weighting factor $\omega(\tilde{x})$ in a demonstrated trajectory with different parameters: (a) with constant ϵ and varying μ, (b) with varying ϵ and constant μ.

4.4 Learning a DMP model

4.4.1 Standard approach

A DMP model is usually learned by human demonstrations. As discussed in [48], the parameters in DMP can be categorized into three types: (1) parameters of the linear term (K and D), (2) parameters in the non-linear term, and (3) the time scaling factor τ and the time constant of the canonical system α.

Given a demonstrated trajectory, the time constant τ and α of the DMP model are usually pre-fixed. τ can be chosen as 1.05 times the duration of the recorded human trajectory to cover the movement fully [48]. Then the time constant α in Eq. (4.2) is adjusted so that the phase variable s converged to zero within τ. The parameters of each basis function ψ (Eq. (4.4)) are also fixed a priori.

In general, it is tricky to learn both the goal attractor and shape attractor of DMP simultaneously. In most literature, the parameters of goal attractor, namely K and D are pre-defined as well. For stability, both matrices must be positive definite. With appropriate ratios of K and D, the system can be critically damped. After pre-specification of all the parameters mentioned above, the goal is then to learn the shape-attraction term f. Given a demonstrated trajectory with M measurements of position, velocity and acceleration $\{x(k), v(k), a(k)\}_{k=1:M}$, with the standard DMP model in Eq. (4.1),

$\boldsymbol{f}$ is determined as follows:

$$\boldsymbol{f}(k) = \operatorname{diag}(\boldsymbol{g} - \boldsymbol{x}_0)^{-1} \left(\tau \boldsymbol{a}(k) - \boldsymbol{K}\tilde{\boldsymbol{x}}(k) + \boldsymbol{D}\boldsymbol{v}(k)\right), \quad k = 1, 2, ..., M. \tag{4.15}$$

By solving the differential equation of the canonical system defined in Eq. (4.2) with initial condition $s(0) = 1$, the phase parameter s is calculated as:

$$s(k) = \exp\left(-\frac{\alpha}{\tau}t_k\right), \quad t_k = (k-1)T_s, \quad k = 1, 2, ...M. \tag{4.16}$$

Then the value of each basis function $\psi(s(k))$ is determined by Eq. (4.4). Together with the values of $\boldsymbol{f}(k)$, the following least square problem is defined:

$$\min_{\Omega} J = (\mathbf{F} - \boldsymbol{\Phi}\boldsymbol{\Omega})^T (\mathbf{F} - \boldsymbol{\Phi}\boldsymbol{\Omega}) \tag{4.17}$$

where

$$\mathbf{F} = \begin{pmatrix} \boldsymbol{f}(1) \\ \vdots \\ \boldsymbol{f}(M) \end{pmatrix} \in \mathbb{R}^{3M \times 1}, \quad \boldsymbol{\Omega} = \begin{pmatrix} \boldsymbol{\Omega}(1) \\ \vdots \\ \boldsymbol{\Omega}(N) \end{pmatrix} \in \mathbb{R}^{3N \times 1},$$

$$\boldsymbol{\Phi} = \begin{pmatrix} \frac{\boldsymbol{\psi}_1(s(1))}{\sum_{j=1}^{N}\boldsymbol{\psi}_j(s(1))}s(1) & \cdots & \frac{\boldsymbol{\psi}_N(s(1))}{\sum_{j=1}^{N}\boldsymbol{\psi}_j(s(1))}s(1) \\ \vdots & & \vdots \\ \frac{\boldsymbol{\psi}_1(s(M))}{\sum_{j=1}^{N}\boldsymbol{\psi}_j(s(M))}s(M) & \cdots & \frac{\boldsymbol{\psi}_N(s(M))}{\sum_{j=1}^{N}\boldsymbol{\psi}_j(s(M))}s(M) \end{pmatrix} \in \mathbb{R}^{3M \times 3N},$$

$$\boldsymbol{\psi}_j = \begin{pmatrix} \psi_{x,j} & 0 & 0 \\ 0 & \psi_{y,j} & 0 \\ 0 & 0 & \psi_{z,j} \end{pmatrix} \quad j = 1, 2, ...N. \tag{4.18}$$

The solution of the optimization problem results in $3N$ parameters that construct $\boldsymbol{f}$. Note that the factor 3 represents three dimensions. As long as the shape attractor is learned, one can use Eq. (4.1) to generate motion trajectories with new parameters. The whole procedure of learning a DMP model is summarized in Fig. 4.4.

4.4.2 Two-step approach

In this thesis, a novel two-step learning procedure is proposed. The design is based on the modified DMP model in Eq. (4.14). Same as the standard approach, several parameters, namely the time constant τ, α as well as μ, ϵ in Eq. (4.11) need to be pre-specified.

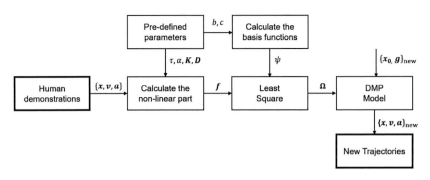

Figure 4.4: Flowchart of the standard learning procedure of a DMP model

The first step aimed to fit a GP model for f. To reduce the influence by the goal-attraction term, the trajectory is segmented, and only the part in which $\omega > 0.9$ is used for calculating f (see Fig. 4.5), in which the weighting of the goal-attraction term is low and has almost no influence on f. It yields:

$$f(k) = \frac{\tau}{\omega(\tilde{x}(k))\gamma} R^{-1} a(k), \quad \forall \omega > 0.9. \tag{4.19}$$

Same as before, the phase parameter s is calculated by Eq. (4.16) and taken as input of the GP model. Then the GP hyperparameters are optimized by maximizing the likelihood using the algorithm developed in [91]. At last, within the concept of sparse GP regression, the mean value μ_u (Eq. (3.24)) and covariance matrix Σ_u (Eq. (3.25)) of the inducing point set X_u are calculated using the SONIG algorithm proposed in [83] and stored as model parameters.

In the second step, the learned GP parameters are used to reproduce the shape attraction part for the whole trajectory, denoted as $\hat{f}$. Then this term is subtracted from the DMP equation so that only the linear goal attraction part is left, namely:

$$\begin{aligned}
f_{lin}(k) &= \frac{1}{1 - \omega(\tilde{x}(k))} \left(\tau a(k) - \omega(\tilde{x}(k))\gamma R \hat{f}(k) \right) \\
&= K(g(k) - x(k)) + D(v_g(k) - v(k)).
\end{aligned} \tag{4.20}$$

The unknown parameters K and D can be identified using the least square method. Fig. 4.6 summarizes the proposed two-step learning procedure. Compared to the standard method, the main advantage is that fewer parameters need to be pre-specified so that the error caused by the wrong parameter choice can be reduced.

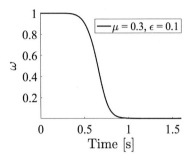

Figure 4.5: The time profile of $\omega\left(\tilde{\boldsymbol{x}}\right)$ in a demonstrated trajectory with $\mu = 0.3\,\mathrm{m}$ and $\epsilon = 0.1\,\mathrm{m}$.

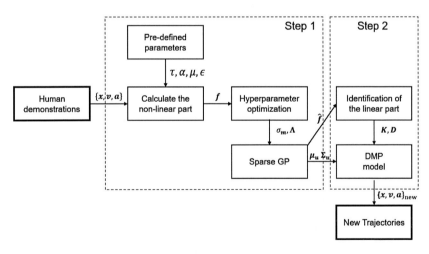

Figure 4.6: Flowchart of the proposed two-step learning procedure of a DMP model

4.5 Evaluation and discussion

This section presents the evaluation of the proposed DMP model based on the human motion data set. Firstly, the human-human handover scenario is considered. During the data collection, the participants were asked to frequently change their initial and handover locations to produce more variations in the motion trajectories. The primary purpose is to validate the advantage of DMP in comparison to purely data-driven models, namely its ability to make predictions with unseen input by the underlying physical knowledge of the system. Then the DMP-GP model is tested in the pick-and-place task and compared with the GP regression introduced in the previous chapter. The aim is to evaluate if the DMP-GP model can also deliver satisfactory results in the scenario where GP works well.

Handover task

The evaluation concerning a human-human handover task was performed as follows: firstly the complete trajectory of one human was shown to the predictor, then it should reconstruct the motion of the other human based on the trained model. In this study, three prediction methods were compared, namely the proposed DMP in this work, the DMP developed in [110] and GP regression with the following definition of input and output :

$$
\mathbf{x} = \begin{pmatrix} \boldsymbol{x}\left(k\right) \\ \boldsymbol{v}\left(k\right) \\ t_k \end{pmatrix} \in \mathbb{R}^{7\times1}, \quad \mathbf{y} = \begin{pmatrix} \boldsymbol{x}'\left(k\right) \\ \boldsymbol{v}'\left(k\right) \end{pmatrix} \in \mathbb{R}^{6\times1}, \tag{4.21}
$$

where $\boldsymbol{x}'$ and $\boldsymbol{v}'$ are the position and velocity of the partner. Thereby, the GP model describes the correlation of motions between the two participants. The prior mean function is set to zero. The time constant τ of both DMP models is set equal to the movement duration. The other parameters are chosen as follows: $\alpha = 10$ (Eq. (4.2)), $\mu = 0.3, \sigma = 0.1$ (Eq. (4.11)).

The overall performance was evaluated by calculating the NRMSE of the prediction regarding the actual measurements. The results are shown in Table 4.1. First of all, the error of GP regression is much larger than that of the two DMP-based methods. As discussed in the previous chapter, it is mainly due to the limitation of a zero-mean GP, in which the prediction only relies on the covariance matrix. If the correlation between test and training data is low, the prediction cannot work well. Although the online GP is able to update the basis distributions with new observations, the adaption is somehow

NRMSE in %	mean	min	max
DMP-GP model in this work	16.34	3.48	38.03
DMP model in [110]	32.13	13.32	62.90
GP regression	62.54	29.12	83.34

Table 4.1: NRMSE of the predictions of human motion with different methods

"delayed". In comparison, the temporal and spatial invariance properties enable DMP to rescale the entire trajectory without retraining the model, which is helpful to deal with the change of goal and duration of the motion.

Comparing the two DMP models, it can be seen that the prediction error with the proposed model in this work is lower than the one in [110]. It is mainly due to the global spatial scaling method presented in Eq. (4.7). As explained in [110], their DMP formalism is derived based on the bio-inspired model in [99], in which the non-linear term is not scaled by initial and goal positions (see Eq. (4.6)). Although the bio-inspired DMP model solves the scaling problem when the start- and endpoint are close, the global scalability is destroyed. It causes new problems, especially when the goal position varies a lot. As shown in Fig. 4.7(a), even though the predicted trajectories have the same form as the training set, their directions are mainly wrong. On the other hand, by using the rotation-based scaling method in [109], the adaptability to new initial- and end positions is improved by keeping the global spatial invariance property. (Fig. 4.7(b)).

Pick-and-place task

The evaluation concerning the pick-and-place task has been done the same as in the previous chapter. Both GP and the proposed DMP-GP model were trained with the same data set. Parameters of the Gaussian error function were changed to $\mu = 0.1, \sigma = 0.025$. The reason is that in this task the total traveling distance is shorter than the handover task. The results of both short and long-term predictions are shown in Fig. 4.8. For short-term prediction, the online sparse GP regression performed sightly better than DMP. The difference is negligible ($< 1\%$). For long-term prediction, DMP delivers better performance. Since DMP is a goal-directed model, its attractive behavior drives the trajectory to the goal position. GP, however, does not have such property.

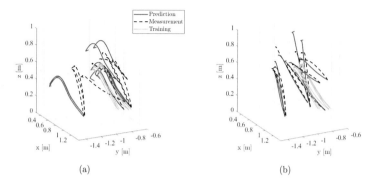

(a)   (b)

Figure 4.7: Results of the predictions of the human 3D paths based on the learned DMP
models: (a) prediction using the bio-inspired DMP model, (b) prediction
using the DMP model in this work. Both subfigures show the same training
set (in grey) and the ground truth (in black).

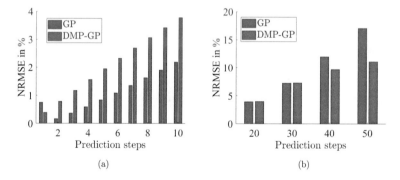

(a)   (b)

Figure 4.8: Performance of the short- and long-term online prediction based on GP regression (blue) and DMP (red) in a pick-and-place task.

Limitations of the DMP-based methods

Although the proposed DMP model outperforms all the other methods studied in this work, i.e., it works well in short- and long-term predictions, and does not require a large training set, and can adapt to goal variations, it still has some limitations. First of all, since DMP has been originally developed for trajectory generation, the goal position and the travelling time are always given. This condition must also be satisfied if one uses DMP for motion prediction. However, this is critical in practice. A possible solution is to develop an algorithm that synchronously estimate the state and parameters, e.g. Extended Kalman Filter ([111]). Still, the convergence takes time and steady errors might occur.

Secondly, to the author's best knowledge, there is still no common conclusion on how the gain matrices of the linear goal attraction term (K, D in Eq. (4.1)) can be learned property. In most literature, these matrices are pre-defined and satisfy three properties: diagonal, positive-definite, and over-damped. However, from the motion prediction point of view, the wrong choice of these parameters introduces error biases and cannot fully be compensated by the shape-attraction term, especially at the end of the motion where the shape-attraction term has almost no influence. Hence, this work aims to identify these gain metrics. The proposed two-step approach is designed for a particular type of DMP formulation, in which the goal and shape-attraction terms are somehow "isolated" through the weighting factor so that they can be learned separately. However, this identification procedure may not work for other DMP formulations. Another issue is that sometimes the identified parameters do not satisfy positive definiteness. It is not critical for motion prediction but will cause problems when using the learned DMP model to generate new motion trajectories since the system is unstable. Hence, it is suggested to define lower bounds for these parameters.

4.6 Coupled DMP for modeling interaction dynamics

The handover scenario discussed in the last section is well-known as a joint action between a giver and a receiver. Both generate motions to drive the relative distance between their hands to zero. This behavior belongs to a special type of collaboration, which aims to modify or keep a task-depending formation. For motion planning and coordination, it is always crucial to find a model that enables describing the interactive behavior between the agents, i.e., the dependency of one's motion on others, especially if no physical contact exists. Computational models such as cooperative optimization [112] or probabilistic primitives [102] are commonly used in literature.

As presented in [48], DMP is originally derived from a second-order system to describe an attractor dynamic. It can be physically interpreted as a damped spring model with a non-linear modulation. The motion is generated by a virtual force between the actual and target position. This interpretation can be extended to formation-oriented collaboration tasks, i.e., each agent is modeled by DMP, and their positions are coupled by the group formation. Within this concept, the interaction dynamic can be described as a coupled mass-spring-damper system with pre-stored potential energy, determined by the initial relative positions between the agents. The non-linear modulation terms in each DMP model can be regarded as an external force. Note that each agent may have its own DMP parameters to generate different dynamic behaviors. The only exception is the time constant τ, which should be kept the same in all DMPs to achieve synchronized motions.

Moreover, it is possible to include the motion control dynamic in the model as well. On the one hand, previous research [113] shows that the human motor control system for performing reaching movements can be represented by mechanical impedance. On the other hand, several Cartesian compliant control strategies (impedance, admittance, and stiffness control) make the robot end-effector also behave like a mass-spring-damper system [52]. As a result, a model with several coupled mass-spring-damper components is obtained, including both the formation and control dynamic.

Considering a multi-agent team with each agent indexed by i and its neighborhood indexed by j. For simplification, it is assumed that the collaboration task is to achieve $\boldsymbol{x}_j - \boldsymbol{x}_i = \boldsymbol{0}$. Based on the DMP formulation in Eq. (4.14), the dynamic model of each agent i is formulated as follows:

$$
\begin{aligned}
\tau \ddot{\boldsymbol{x}}_{i,d} &= (1 - \omega_{ij}) \left(\boldsymbol{K}_i \left(\boldsymbol{x}_j - \boldsymbol{x}_i \right) + \boldsymbol{D}_i \left(\dot{\boldsymbol{x}}_j - \dot{\boldsymbol{x}}_i \right) \right) + \omega_{ij} \gamma_i \boldsymbol{R}_i \boldsymbol{f}_i(s), \\
\boldsymbol{M}_{c,i} \ddot{\boldsymbol{x}}_i &= \boldsymbol{K}_{c,i} \left(\boldsymbol{x}_{i,d} - \boldsymbol{x}_i \right) + \boldsymbol{D}_{c,i} \left(\dot{\boldsymbol{x}}_{i,d} - \dot{\boldsymbol{x}}_i \right) + \boldsymbol{M}_{c,i} \ddot{\boldsymbol{x}}_{d,i},
\end{aligned}
\tag{4.22}
$$

where $\boldsymbol{M}_{c,i}, \boldsymbol{D}_{c,i}, \boldsymbol{K}_{c,i}$ are diagonal control gains defined by the impedance control loop. $\boldsymbol{x}_{i,d}$ is the position reference for the control loop in agent i, s is the common phase variable defined in Eq. (4.2). The weighting factor ω_{ij} is determined as follows:

$$
\omega_{ij} = 0.5 \left(1 + \mathrm{erf} \left(\frac{\| \boldsymbol{x}_j - \boldsymbol{x}_i \| - \mu}{\sqrt{2} \epsilon} \right) \right).
\tag{4.23}
$$

The model shows that the motion of each agent i is driven by the sum of three virtual forces. The first one is generated by the spring-damper system with parameters $\boldsymbol{K}_i, \boldsymbol{D}_i$, which aims to shorten the relative distance to the neighborhood. The second one is generated by the shape-attraction term in the DMP model, which aims to modulate the dynamic motion. The third one is generated by the motion control loop of each agent i, which can be also represented by a spring-damper system with parameters $\boldsymbol{K}_{c,i}, \boldsymbol{D}_{c,i}$

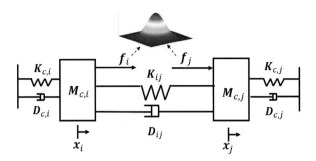

Figure 4.9: Equivalent model of a formation-orient collaboration task using DMP formalism.

and aims to suppress the control error. A graphical illustration is shown in Fig. 4.9, in which $\boldsymbol{K}_{ij} = \frac{1-\omega_{ij}}{\tau}\left(\boldsymbol{K}_i + \boldsymbol{K}_j\right), \boldsymbol{D}_{ij} = \frac{1-\omega_{ij}}{\tau}\left(\boldsymbol{D}_i + \boldsymbol{D}_j\right)$.

Note that for system passivity all the gain matrices must be positive definite and the weighting factor must satisfy $1 - w_i > 0, \forall i$. The system's nonlinearities exist in the state-dependent scaling and weighting factors. In case that the task configurations remain invariant, the model can be further simplified by setting the scaling factors to constant values.

Moreover, the model can be extended by considering the correlations in the non-linear modulation term in each agent's DMP. Since they are modeled as GP, it is straightforward to correlate them with their joint distribution:

$$\begin{pmatrix} \boldsymbol{f}_i \\ \boldsymbol{f}_j \end{pmatrix} \sim \mathcal{N}\left(\begin{pmatrix} \boldsymbol{m}_i \\ \boldsymbol{m}_j \end{pmatrix}, \begin{pmatrix} \boldsymbol{k}_i & \boldsymbol{k}_{ij} \\ \boldsymbol{k}_{ij}^T & \boldsymbol{k}_j \end{pmatrix} \right), \qquad (4.24)$$

where $\boldsymbol{m}$ and $\boldsymbol{k}$ represent mean and covariance function respectively. Thereby, agent i can estimate the forcing term of its neighbor j by calculating the conditional probability $p\left(\boldsymbol{f}_j | \boldsymbol{f}_i\right)$, namely:

$$\bar{\boldsymbol{f}}_j = \boldsymbol{m}_j + \boldsymbol{k}_{ij}^T \boldsymbol{k}_i^{-1}\left(\boldsymbol{f}_i - \boldsymbol{m}_i\right), \qquad (4.25)$$
$$\mathrm{cov}\left(\boldsymbol{f}_j\right) = \boldsymbol{k}_j - \boldsymbol{k}_{ij}^T \boldsymbol{k}_i^{-1} \boldsymbol{k}_{ij}. \qquad (4.26)$$

Now the interactive behavior can be separated into two terms: a linear state-dependent term, represented by the two coupled mass-spring-damper systems and a state-independent probabilistic term, represented by the joint distribution of two GPs. Both terms can be learned separately.

To examine the equilibrium of the system, all the time derivatives as well as the phase variable s are set zero. It yields:

$$\boldsymbol{K}_{c,i}\left(\boldsymbol{x}_{i,d,\infty} - \boldsymbol{x}_{i,\infty}\right) = \boldsymbol{0} \tag{4.27}$$

$$\boldsymbol{x}_{i,\infty} - \boldsymbol{x}_{j,\infty} = \frac{\omega_{ij}}{1 - \omega_{ij}}\left(\boldsymbol{K}_i - \boldsymbol{K}_j\right)^{-1}\left(\gamma_i \boldsymbol{R}_i \boldsymbol{f}_i(0) + \gamma_j \boldsymbol{R}_j \boldsymbol{f}_j(0)\right). \tag{4.28}$$

One sufficient condition to achieve the desired equilibrium $\boldsymbol{x}_{i,\infty} = \boldsymbol{x}_{i,d,\infty} = \boldsymbol{x}_{j,\infty}$ is that $\boldsymbol{f}_i(0), \boldsymbol{f}_j(0) \to \boldsymbol{0}$. However, this condition is not always satisfied if $\boldsymbol{f}_i$ and $\boldsymbol{f}_j$ are modeled by GP as in Eq. (4.5). On the other hand, the weighting factor $\omega_{ij}\left(\boldsymbol{x}_j - \boldsymbol{x}_i\right)$ defined in Eq. (4.11) only converges to zero if $\boldsymbol{x}_{r,\infty} = \boldsymbol{x}_{h,\infty}$. Consequently, if $\omega_{ij} \neq 0$ and $\boldsymbol{f}_i(0), \boldsymbol{f}_j(0) \neq \boldsymbol{0}$, the relative position between agent i and agent j will converge to a undesired value, namely $\boldsymbol{x}_{j,\infty} - \boldsymbol{x}_{i\infty} \neq \boldsymbol{0}$.

A practical solution to the problem is to force the the weighting factor ω_{ij} to converge to zero. Defining a new variable $\tilde{\boldsymbol{x}}_{ij}$ for ω_{ij} as follows:

$$\tilde{\boldsymbol{x}}_{ij} = \min\left(\Delta\boldsymbol{x}_{ij}, \Delta\boldsymbol{x}_{ij,n}\right), \tag{4.29}$$

where

$$\Delta\boldsymbol{x}_{ij} = \|\boldsymbol{x}_j - \boldsymbol{x}_i\|, \tag{4.30}$$

$$\Delta\boldsymbol{x}_{ij,n} = \|\boldsymbol{x}_{j,0} - \boldsymbol{x}_{i,0}\| \exp\left(-\beta t\right). \tag{4.31}$$

$\Delta\boldsymbol{x}_{ij}$ represents the actual distance between agent i and j. $\Delta\boldsymbol{x}_{ij,n}$ describes an artificial convergent behavior. It starts with the initial distance and converges to zero with a positive time constant β. The value of β is tuned in the way that $\Delta\boldsymbol{x}_{ij}(t) < \Delta\boldsymbol{x}_{ij,n}(t)$ holds in most cases. Only if $\Delta\boldsymbol{x}_{ij}(t \to \infty) \neq 0$, the convergence of $\Delta\boldsymbol{x}_{ij,n}$ will force $\omega(\tilde{\boldsymbol{x}})$ converge to zero. Then the right side of Eq. (4.28) is zero so that $\boldsymbol{x}_{i,\infty} - \boldsymbol{x}_{j,\infty} = \boldsymbol{0}$.

The model can be further generalized to varies of cooperation tasks with attractive (e.g. formation maintenance) or repulsive (e.g. obstacle avoidance) behavior, depending on how energy flow in the system is defined. It can be used for simulating the interactive behavior between agents, providing quantitative information for model-based control design methods.

4.7 Summary

In this chapter, human motion prediction based on dynamic movement primitives was studied. DMP is characterized as a hybrid physical and data-driven approach to describe goal-directed behavior, combining a simple linear dynamic model with statistical learning techniques. To overcome several limitations in the conventional DMP formulation, a modified DMP model was presented, including a GP-based shape learning component, a factor to adjust weightings of the linear and nonlinear terms, and a rotation-based formulation to maintain the global scalability. Simulation results show that DMP outperforms almost all the other methods that have been implemented so far in this work. It works well in short- and long-term predictions, does not require a large training set, and is robust against variation of initial and goal positions. Moreover, the DMP model can be extended to describe interactive dynamics between multiple agents. On the one hand, their kinematic values can be coupled through the linear part of DMP, which is physically interpreted as a virtual spring-damper force. On the other hand, properties of GP enable coupling of their shape attraction terms with joint distribution.

However, the proposed DMP formulation has several limitations and problems as well. First of all, motion prediction based on DMP always requires a pre-specified target position and time constant that influences the convergence rate of the canonical system. If both of them are unknown, no prediction can be made. One possible solution is to design an algorithm that enables online state and parameter estimation, which can be challenging in practice, especially when the model is nonlinear or the parameters are time-variant. Secondly, the learned parameters in the goal-attraction term were sometimes physically infeasible, i.e., the positive definiteness of the gain matrices cannot always be satisfied. From the identification of view, it might be caused by the error in the other terms and does not significantly influence the prediction results. However, problems may occur when using the learned model to generate more trajectories since the convergence property of DMP is destroyed. At last, the proposed DMP is specially designed for translational movement. For rotational motion, the model needs to be further adjusted.

In recent years, DMP has been intensively studied and integrated in various research areas in human-robot collaboration, such as cooperative manipulation, impedance learning, reinforcement learning, human motion recognition, etc. [97] Later in Chapter 8, a DMP based adaptive learning and control framework with application to human-robot handovers is presented.

5 Discussion on Human Motion Prediction for Human-Robot Collaboration

5.1 Comparison of different approaches in human motion prediction

The objective of the last three chapters is to develop an approach for human motion prediction that can be applied easily for online robot trajectory planning and control in the context of human-robot collaborations. Totally five different methods, (1) Kalman filtering with constant acceleration model, (2) polynomial regression with minimum jerk model, (3) inverse optimal control, (4) online sparse Gaussian Process regression, and (5) Dynamic Movement Primitives with Gaussian Process, have been studied, implemented, and evaluated with the help of human motion data sets. From the modeling point of view, these methods can be categorized into physical-based (1,2), data-based(3,4), and hybrid approaches (5). From the methodological point of view, they can be classified into goal-oriented[1] (2,3,5) and state-oriented approaches (1,4). The former relies on a given set of goals and aims to predict how humans may reach these goals. The latter focuses on the update rule that describes how states evolve. A more general discussion is made in this chapter to understand better how these methods address different practical needs and which performance level they can achieve.

5.1.1 Boundary conditions

At first, the boundary conditions for the analysis are summarized as follows. All the methods discussed in this chapter are constrained by human hand-reaching movements. Only the Cartesian translational motion of three DOFs (XYZ) is considered. The environment is static and obstacle-free. The workspace is not beyond the reach of the

---

[1]It is also named as the planning-based approach in other literature.

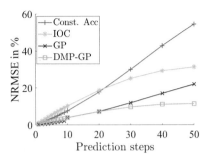

Figure 5.1: Normalized mean prediction error of all the methods discussed in this work under different prediction horizons.

human arm. In other words, all the motions are provided within the human kinematic constraints. The travelling distance and duration are therefore short (within 0.5 m and 2 s). Some variances are added in the task configurations by varying the initial and target positions. Both stand-alone and dyadic motions are included.

The human hand position was recorded by an optical motion capture system with high precision (0.5 mm) and low noise level. Hence, the error caused by the measurement process can be neglected. Velocity and acceleration were estimated based on the position values.

The validity of the discussions remains within the boundary conditions listed above, which is a potential limitation of this thesis. Yet, some generalizable conclusions can be drawn by comparing the results with several related works.

5.1.2 Prediction accuracy at different time scales

Fig. 5.1 plots the mean prediction error of all the five methods as a function of prediction horizon in a pick-and-place scenario. Note that the comparison is made in an ideal situation, in which the goal positions and total traveling time are known to all predictors. Later the required prior information for each predictor will be discussed separately.

First of all, for short prediction horizons (< 10 or 0.1 s), Kalman filtering (KF) with a simple kinematic model (constant acceleration) has already delivered satisfying performance (normalized error $< 10\%$). Other popular forms include the constant velocity and the coordinated turn model [114] (for curvilinear motion). These models are usually formulated by a recurrence equation with the assumption that the nth-order differentiation between two position samples (depending on the definition) remains zero, and the approximation error is white noise. Such models are suitable to predict motions with fewer uncertainties in a static environment for a short look-ahead time. One related work [16] shows that this approach can also be implemented for the prediction of rotational motion.

To achieve a long-term prediction or consider a dynamic environment (e.g., obstacles), simple kinematic models with local approximations are insufficient. It is necessary to include more information in the prediction model, especially the motion generation mechanism in the human motor control system. The most commonly used approach is to describe human control behavior as an optimization problem. Even these models are not directly derived from physical laws but consist of the analytical solutions of an optimization problem, they are still classified as physical-based approaches in this thesis since the cost functions in the optimization problem are defined from a physical point of view. As presented in Chapter 2, motion prediction with minimum jerk model is based on its analytical solution, namely a fifth-order polynomial function. The results in this work show that this model only works when its coefficients are all known. Hence, it becomes more like a trajectory fitting problem rather than an online prediction. However, results in Chapter 2 show that simultaneous state and parameter estimation can be problematic.

Next, inverse optimal control has been implemented as an extension of the minimum jerk model. Its cost function contains multiple optimization criteria, and the purpose is to identify the weightings of each criterion. In this thesis, it is classified as a data-driven approach since solving an inverse optimization problem requires a data set of human demonstrations. As long as the cost function is determined, one can solve the optimization problem forwardly to predict human motion. Its performance is better for short-term prediction than the minimum jerk model but not for long-term prediction. Like the minimum jerk model, it also requires prior information on the task but the implementation is more complex.

So far, it can be seen that even for a simple human hand reaching motion, it is not trivial to find a simple model or cost function that can fully explain the motion dynamic and work for both short and long-term predictions. Complex biomechanical models considering human kinetics are beyond the scope of this thesis due to their high computational load and sensory requirements. On the other hand, non-parametric approaches, in which no specification of the structure of the model is needed, might be an option. In this work, Gaussian Process (GP) is used to describe transitions between the current state (position and velocity) and the next one. Moreover, the distribution is recursively updated by incoming measurements (online GP). Fig. 5.1 shows that the prediction works well for both short- and long predictions. However, one disadvantage is that it requires a training set that includes all motion types that need to be predicted, making the model less flexible and generalizable.

To overcome that issue, one needs to think back to physical models. However, the aim is not to fully cover the motion dynamics but rather interpret some "basic" characteristics, for instance, the "attractive" behavior from the initial to target position of a reaching movement. Besides, the residual is modeled by machine learning methods. This concept is categorized as a hybrid physical and data-driven approach. Dynamic Movement Primitives (DMP) is one of the best examples. In this work, DMP is combined with GP to utilize its strength in approximating nonlinear functions. Fig. 5.1 shows that the DMP-GP method provides the best overall performance in comparison to other methods.

Another important factor is the error growth rate concerning the prediction horizon. In Fig. 5.1 one can see a significant difference between state-oriented (KF,GP) and goal-oriented (DMP, IOC, minimum jerk) approaches. As introduced at the beginning of this chapter, state-oriented approaches focus on local transitions between states. Results in this work show that the error increases almost linear to prediction horizons. On the other hand, the error growth rate of goal-oriented approaches is saturated with prediction horizons. It is logical since the uncertainties become smaller as the goal has been reached. Hence, goal-oriented approaches are more beneficial for long-term predictions.

Comparison with related work

Results in this thesis are compared with results from two related publications [115, 116] with similar concepts. They both study human hand reaching motion predictions and compare different physical-based and data-driven approaches. Fig. 5.2(a) illustrates the experimental setup in [115]. During this task, human participants picked and placed several screws at designated locations on a table while the robot moved a brush over the screws to simulate sealant work. The human hand movements were tracked by a motion capture system. The purpose is to evaluate different approaches to human motion prediction, including (1) physical-based method (constant velocity model), (2) data-driven method (time series classification and sequence prediction), (3) multi-predictor method with a fusion of the methods above. Fig. 5.2(b) shows the experimental setups in [116]. This work aims to predict human hand movement trajectories to grab test tubes with different target positions. The human motion was captured by a Kinect depth sensor and only upper body joints were considered. Totally four approaches have been evaluated, including (1) GP regression, (2) minimum jerk model, (3) Bezier curve

---

[1]Video source: https://www.youtube.com/watch?v=Dk5XVQBDJpU Last viewed: 15.02.2022

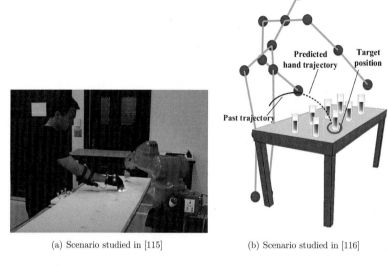

(a) Scenario studied in [115]   (b) Scenario studied in [116]

Figure 5.2: Graphical illustration of the experimental setups in two related publications.

model, (4) combination of classification and regression.

Fig. 5.3 illustrates results pf human motion prediction in [115, 116]. Note that a direct comparison of the values in Fig. 5.3 with Fig. 5.1 does not make much sense since the experimental conditions and software tools are different. Still, by analyzing the tendencies, some common ground can be found.

As shown in Fig. 5.3(a), the prediction error of the "velocity-based" (constant velocity model) method has the same characteristics as the constant acceleration model implemented in this work, namely: (1) low error on short-term prediction, (2) error increases linearly for long look-ahead time. The "time series classification method" and "sequence prediction method" belong to data-driven approaches and perform better for long-term prediction, matching the results in this work. The difference is that their error for short look-ahead time is relatively high. The possible reason is that the prediction is based on the classification results, which might be less reliable in the early motion phase. Similar to this work, a predictor with a fusion of physical and data-driven methods provides the best results.

In another related publication [116], two of four methods are the same as in this work,

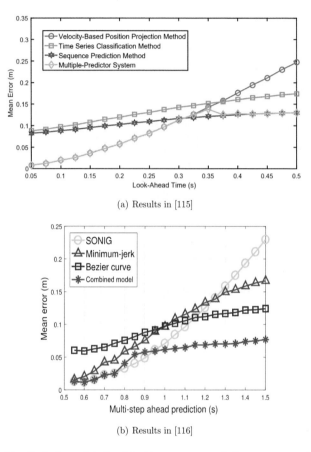

(a) Results in [115]

(b) Results in [116]

Figure 5.3: Results in two related publications: mean prediction errors as a function of look-ahead time.

namely the minimum jerk model and online GP regression (SONIG). The result of online GP as a state-oriented approach is similar. However, the minimum jerk model delivers a much better performance comparing to the results in this work. The possible reason is that instead of a fifth-order polynomial function, the authors use another mathematical formulation derived in [73], which consists of three independent exponential functions. On the other hand, the authors also point out that inaccurate parameter identification can cause large error in long-term predictions. The "Bezier curve" is a polynomial function parameterized by several points on the path [117], including starting and goal positions. The so-called "combined model" is a data-driven approach that combines time series classification of motions trajectories (depending on the goal position) with Bayesian regression. As illustrated in Fig. 5.3(b), the error growth rate of the three goal-oriented approaches is consistent with results in this work, showing their advantages in long-term prediction.

5.1.3 Implementation complexity and scope of application

Table 5.1 briefly summarizes the implementation complexity and possible applications of all the methods discussed in this work. Motion prediction based on the Kalman filter with simple physical models has the lowest implementation complexity. It does not require any task or environment information. Only the parameters of the Kalman filter (process and measurement noise) need to be pre-specified. This method is suitable for motion tracking or short-term predictions.

Linear regression with polynomial functions is more suitable for trajectory fitting or smoothing. It usually requires pre-specification of the order of the function, initial and terminal states, and the total time. The computational complexity is low. This method can also be used as a motion pattern for trajectory planning.

GP regression needs an off-line training phase with several recorded human demonstrations. Pre-specified parameters are its inputs and outputs, the form of the kernel function (e,g, squared exponential, dot product, etc.), and the prior mean function (usually set to zero). The training phase optimizes the hyperparameters and computes the posterior distributions of the training set as "basis vectors" for online prediction. GP has a high computational complexity ($\mathcal{O}\left(M^3\right)$ for M data points) but that can be reduced through sparse representation. It is unnecessary to include any task/environment information in a GP model since it mainly works on distributions. However, as long as the task or environment varies, the model needs to be retrained. GP regression works for both short and long-term predictions and can deal with uncertainties.

The proposed DMP-GP method can be regarded as an extension of the GP regression

Method	Comlexity	Offline training	Required information	Applications
KF with const. acc.	low	no	process- & measurement noise for KF	target tracking, state estimation short-term prediction
Min. jerk	low	no	target position, total time	trejectory filtering, smoothing, planning
Online GP	high	yes	kernel function, inputs & outputs, prior mean function	uncertainty-aware online motion predictions &, trajectory planning
DMP-GP	high	yes	target position, time constant, parameters for GP (see above), parameters for DMP (gain matrices)	online motion predictions, high-DOF trajectory planning, control, imitation learning
IOC	high	yes	system model, target position, excution time, features in the cost function	online motion predictions, imitation learning, learning objective functions

Table 5.1: Implementation complexity and applications of all the methods discussed in this thesis for prediction of human reaching movements

method. It contains more parameters that need to be pre-defined, namely the target position, the time constant, the Gaussian error function in the weighting factor, and the gain matrices in the goal-attraction term (see Fig. 4.6). In literature, these parameters are either chosen by trial-and-error or identified from demonstrated trajectories. An offline training phase is also required to determine the GP model. Although the model is more complex than a zero-mean GP, the online computational complexity remains almost the same since most of the computational load lies on the GP term. DMP also works for short and long-term predictions, is flexible to task reconfiguration, and can easily be combined with trajectory planning and control.

Motion prediction based on inverse optimal control (IOC) has a high implementation complexity. It requires solving two constrained optimization problems: a maximum entropy approach to identify the cost function (offline) and optimal control (online) to generate predicted motion trajectories. Both optimization problems require a system model, which can be nonlinear and make the solution even more complex. Critical task information such as target position and total execution time is also required for online prediction. This method works for both short and long-term predictions. But the results in this work show that there is no significant improvement in the performance. Even though it is still a valuable tool to help better understanding the human control strategy, and can be combined with other learning-based approaches.

5.2 Potential research trends

5.2.1 From single-predictor to multiple-predictor

Results in this work as well as in [115, 116] suggest that a combination of different modeling/prediction approaches will enhance the performance by utilizing the strengths of every single method (see Fig. 5.1 and Fig. 5.3. Moreover, two recent surveys on human modeling [14] and human motion trajectory prediction [15] have drawn the same conclusion. Hence, there is a reason to believe that a multiple-predictor framework will be a future research trend.

There are various possible concepts for a multiple-predictor design. For instance, authors in [115] proposed a multiple-predictor system with a "parallel" structure, which is graphically illustrated in Fig. 5.4(a). Every single predictor is trained offline individually to learn its parameters. In the online phase, all predictors run parallel, and their outputs are compared to the sensor feedback. According to the errors, a weighting factor for each predictor is calculated. Then a model selector decides which method should be used at

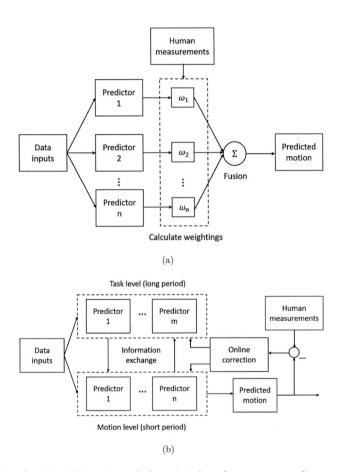

Figure 5.4: Graphical illustration of the principles of two concepts for a multiple-predictor framework: (a) parallel structure (b) hierarchical structure.

which look-ahead time. Although the results show that such a multiple-predictor system improves the robustness and generalizability for various scenarios, it is argued that the concept is not ideal. Especially the following two questions still remain: (1) How to make up for one predictor's deficiencies by using other's strong points? (2) How to make corrections on the parameters of each predictor based on measurements?

With consideration of these questions, this thesis suggests a hierarchical structure for a multi-predictor framework (Fig. 5.4(b)). Firstly, the predictors are categorized into two groups: task-level and motion-level. The task-level predictors aim to decide which task or type of motion the human is currently executing and pass the information to motion-level predictors. Methods such as time series classification or inverse optimal control can be used for task-level predictors. They usually require multiple samples of human motion trajectories to gather enough inputs for prediction so that the running period is relatively long. The motion-level predictors (e.g., DMP, GP, etc.) take the task information and incoming human measurements to predict future motion states. Moreover, the predicted values are sent back to the task-level predictors as additional inputs to help them readjust their predictions. For instance, the classification accuracy is usually proportional to the percentage of the total trajectory that is observed. In the early phase, in which only a small number of measurements are available, the outputs of motion-level predictors can be used as "virtual" inputs for task-level prediction. In summary, the information exchange between different predictors will play an essential role in performance improvement. Secondly, it is also important to take advantage of human-motion feedback to recursively correct/learn the model parameters, making the framework adaptive and robust to dynamic environment and human partners.

5.2.2 From simple motions to complex skills

Many related works, including this thesis, mainly focus on studying one particular type of human motion: object handover/take over, grasping, transporting, etc. Robots should learn and take over more specialized and complex manipulation tasks to achieve seamless and intelligent human-robot collaborations. It is believed that there are two central issues for this purpose: (1) representation and abstraction of human skills, (2) scalable learning approaches.

Results in this work suggest that a hybrid physical and data-driven approach with the concept of movement primitives can be a possible solution for the first problem. On the one hand, physical characteristics such as kinematic/dynamic constraints should be taken as essential "building blocks" in representing and reasoning manipulation tasks [118, 119]. Based on such building blocks and demonstrated trajectories, a "human motion library" can be built, containing various "task primitives" that compose manipulation skills. On the other hand, data-driven/probabilistic approaches are responsible for combining, classifying, and learning the task primitives [120, 121]. Moreover, a data-driven model is also a valuable tool to describe features that are impossible to model physically. Last but not least, to reproduce/execute the learned skills, stochastic optimization plays an essential role for robot motion planning in awareness of uncertainties [122, 123].

A manipulation task usually includes a series of movements involving multiple DOFs of the human body and robot. It might result in a high-dimensional feature space and provides challenges in the learning progress. For example, GP suffers from cubic complexity to data size. Scalable learning approaches with sparse and local approximations show a perspective of dealing with large-scale learning problems [124] and have drawn more and more attention in machine learning communities.

5.2.3 From individual to collaborative behaviors

Learning individual behaviors is the first step to achieve a successful collaboration. In the next stage, it is crucial to develop high-level cognitive mechanisms to feel, evaluate and understand the partner's intentions and actions [6]. In Chapter 4, a preliminary design for modeling the interactive dynamics is presented, which considers coupling effects in both physical and latent states. However, it is still a low-level and rough description of collaborative behaviors. Modeling a joint action has also been intensively studied in other subjects, e.g., social science, psychology, cognitive science, etc. However, many of their models cannot be directly integrated into robot control design since they are usually not mathematically formulated. Several early works show that methods such as symbolic representation [125], state machines [126] or game theory [127] are possible candidates. Bridging the gaps between different subjects and developing a model that comprehensively describes the collaborative behavior and can be easily implemented and integrated into robot control would be a topic with high potential.

6 Physical Human-Robot Collaboration and Impedance Control

6.1 Introduction

While previous chapters were mainly concerned with human motion prediction, this chapter addresses robot control and focuses on the physical human-robot collaboration (PHRC). As defined in [1], in PHRC, "human(s), robot(s) and the environment come to contact with each other and form a tightly coupled dynamical system to accomplish a task". In the author's view, it is more challenging for control design when physical interaction between human and robot arises. First of all, the generated motion/force by the robot will directly act on the human via contact force, which is safety-critical and needs to be handled carefully. Secondly, the human and the robot have to agree in moving direction and share the load which requires effective coordination/negotiation strategies. Hence, conventional motion control approaches, which only focus on tracking a reference trajectory in free space, are insufficient. The purpose of this thesis is to develop an advanced control strategy that enables the robot cognitively interact with human and makes proactive contribution to the task.

Following the standard control design procedure, this chapter starts with the analysis of the system dynamics. Since multiple agents are involved in the collaboration, building a model that includes all the characteristic features is critical, especially an explicit description of the interaction forces/torques. At the same time, it should remain simple and compact. This chapter analyses the collaboration task based on the model proposed in [128] which strictly focuses on the interaction effects of manipulators and objects in task space. One of the key aspects is to simplify the description of individual agent dynamics based on the principle of impedance control.

Impedance control [54] is a widely known interaction control approach. The key concept is to modulate the robot dynamic as mechanical impedance. Impedance control has been intensively studied in literature from the control perspective. However, when regarding it as a part of the system model, the equivalence of the apparent impedance dynamic to its desired value needs to be verified. This chapter reviews several common forms of impedance control with special focus on possible coupling effects in the resulted impedance behavior.

After analyzing the physical characteristics of the collaboration, the next question is how to formulate the task from both the physical and cognitive point of view. Game theory has been shown to be suitable for describing various interaction behaviors, depending on the definition of the cost function and task constraints [129, 4]. In this work, the human-robot collaboration is formulated as a differential game, and the solution is defined by Nash equilibrium.

The rest of this chapter is organized as follows. Section 6.2 briefly reviews the most commonly implemented impedance control laws with a special focus on the comparison of their resulted impedance behaviors. Section 6.3 firstly analyzes the collaboration task from both the physical constraints and impedance control point of view, then discusses possible problem formulations of a collaboration task. As usual, Section 6.4 summarizes the most important points.

6.2 Impedance control

6.2.1 Classification of robot interaction control

In physical human-robot collaboration, the robot continuously interacts with the environment (including humans) through contact force acting on its end-effector. Hence, robot interaction control's key issue is to deal with motion tracking and contact force properly. Fig. 6.1 presents an overview of the most commonly used interaction control methods in the literature. They can be categorized into two main groups: direct force control and indirect force control. The former considers the contact force the controlled output and requires a force feedback control loop. The latter aims to achieve force control via motion control. The Hybrid force/position control separates motion and force control into two decoupled sub-problems, and executes them in complementary frames. It is still classified as direct force control since a force feedback control loop is required [52].

In many collaboration tasks, such as transporting a bulky object or assembling a product, it is difficult to define a consistent reference value for the contact force. It relies on the task configuration, the contact surface, and even the behavior of the human partner. Hence, the indirect force control strategy is more frequently used in physical human-robot collaboration. Although contact force measurements are unnecessary, they can

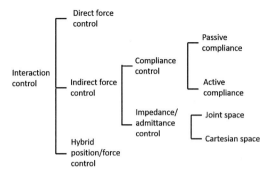

Figure 6.1: An overview of interaction control methods in robotics

still be used as a feedback signal to evaluate the collaboration performance and learn optimal control parameters [130].

Indirect force control uses a motion control strategy to influence the contact force. The critical issue is to define some criteria to achieve a reasonable compromise between precise tracking of motion reference and suppressing contact force. These criteria result in a "compliant" behavior of the robot. Compliance control and impedance/admittance control are the most commonly implemented techniques. Compliance control concentrates on the static relationship between the end-effector pose derivation and the contact force. It can be further classified as passive and active compliance. The former is achieved by mechanical devices, e.g., the remote center of compliance (RCC) [131], the latter is realized by proper design of the control gains. Impedance/admittance control attempts to regulate the dynamic relationship between the end-effector and contact force, which is described by a second-order dynamic system consists of mechanical impedance (mass, spring, and damper) [54]. From the control engineering point of view, their main difference lies in the control structure. Impedance control contains an inner torque control loop, while admittance control has an inner motion control loop [52]. Impedance/admittance control can be designed and executed in both joint and Cartesian space.

This thesis concentrates on impedance control. Note that since impedance control is actually motion control, it can also be applied for trajectory tracking in contact-free tasks. This property makes it possible to build a general framework upon the standard impedance control structure, which can work for various collaboration tasks with or without physical interaction. The rest of this section introduces the principle and formulation of impedance control in both joint and Cartesian space.

6.2.2 Joint impedance control

Considering the following robot dynamic equation:

$$M\left(q\right)\ddot{q} + C\left(\dot{q}, q\right)\dot{q} + g\left(q\right) = \tau - \tau_{ext}, \tag{6.1}$$

where q represents joint angles, M is the joint inertia matrix, C is the matrix describing centrifugal and Coriolis effects, g is the vector of gravitational torques, τ and τ_{ext} are the joint torque generated by motors and the contact force respectively.

Assuming that the robot control task is to follow a pre-defined joint motion trajectory, denoted as $(\ddot{q}_d, \dot{q}_d, q_d)$, consider the following control law:

$$\begin{aligned} \tau = & M\left(q\right)\ddot{q}_d + C\left(\dot{q}, q\right)\dot{q} + g\left(q\right) \\ & + M\left(q\right) M_d^{-1}\left(D_d\dot{\tilde{q}} + K_d\tilde{q}\right) + \left(I - M\left(q\right) M_d^{-1}\right)\tau_{ext}, \end{aligned} \tag{6.2}$$

where $\tilde{q} = q_d - q$ represents the joint position error, $M_d, K_d, D_d \succ 0$ are control gains. Substituting Eq. (6.2) into Eq. (6.1), it yields the following joint control error dynamic:

$$M_d\ddot{\tilde{q}} + D_d\dot{\tilde{q}} + K_d\tilde{q} = \tau_{ext}, \tag{6.3}$$

which is equivalent to the differential equation of a second-order mechanical system. M_d, K_d, D_d correspond to mass, spring and damper respectively. Briefly speaking, the purpose of joint impedance control is to make each joint behave like a mass-spring-damper system in reaction to the external torques (see Fig. 6.2(a)). In other words, impedance control aims to achieve a desired relationship between the motion deriva-tion and the external torques rather than directly control their values. To keep the joint impedance decoupled during the interaction, the control gains must be diagonal. Furthermore, it is necessary to measure the external torques in each joint.

Other forms

If the torque measurements are not available, one can also exclude the torque-related term, then the control input becomes:

$$\tau = M\left(q\right)\ddot{q}_d + C\left(\dot{q}, q\right)\dot{q} + g\left(q\right) + M\left(q\right) M_d^{-1}\left(D_d\dot{\tilde{q}} + K_d\tilde{q}\right) \tag{6.4}$$

---

[1]Image source: https://www.nist.gov/hahhadin-johannesmeierpdf

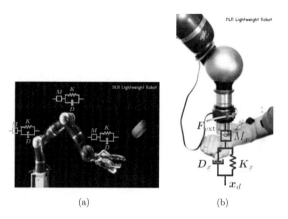

Figure 6.2: Graphical illustration of the principle of impedance control: (a) joint space, (b) Cartesian space.[1]

which results in the following impedance behavior:

$$M_d\ddot{\tilde{q}} + D_d\dot{\tilde{q}} + K_d\tilde{q} = M_d M^{-1}(q)\,\tau_{ext}. \tag{6.5}$$

The presence of M^{-1} makes the system coupled.

Another variation is to choose $M_d = M^{-1}$ so that the last term in Eq. (6.2) vanishes. Therefore, the feedback of external torques can be avoided. This method is named as "impedance control without inertial shaping" [132] since M_d cannot be designed arbitrarily. Under its concept, the joint error dynamic becomes:

$$M(q)\ddot{\tilde{q}} + D_d\dot{\tilde{q}} + K_d\tilde{q} = \tau_{ext} \tag{6.6}$$

The coupling effect now exists in the first term. If the joint accelerations are small, which is normally the case when robot interacts with human, this term can be neglected.

Moreover, it is possible to realize impedance behavior in joints through PD control with gravity compensation, a widely used joint position control strategy. The joint control input is defined as:

$$\tau = C(\dot{q}, q)\,\dot{q} + g(q) - D_d\dot{q} + K_d\tilde{q}, \tag{6.7}$$

where $\ddot{q}_d, \dot{q}_d$ are set zero, and the inertial related terms M_d, M are neglected. The resulted joint dynamic is described by:

$$-M(q)\ddot{q} - D_d\dot{q} + K_d\tilde{q} = \tau_{ext}. \tag{6.8}$$

If the robot's motion is slow, i.e, $\ddot{q}_d, \dot{q}_d \approx 0$, the system can be regarded as a set of decoupled springs. Hence, some literature classifies it as joint compliance control. If there is no specific requirement on the equivalence of the reproduced impedance system, this method can also be considered due to its simplicity in implementation.

6.2.3 Cartesian impedance control

Cartesian impedance control is of great practical importance since the physical interaction between robot and environment happens mostly in Cartesian space. Its principle is similar to joint impedance control, namely to make the robot end-effector behave like a mass-spring-damper system described in Cartesian space (see Fig. 6.2(b)). Denoting the end-effector reference trajectory as $(\ddot{x}_d, \dot{x}_d, x_d)$, the control input is defined as:

$$
\begin{aligned}
\tau =& M(q)J^{-1}(q)M_d^{-1}\left(M_d\ddot{x}_d + K_d\dot{\tilde{x}} + D_d\tilde{x} - M_d\dot{J}(\dot{q}, q)\dot{q}\right) \\
&+ \left(J^T(q) - M(q)J^{-1}(q)M_d^{-1}\right)h_{ext} \\
&+ g(q) + C(q, \dot{q})\dot{q},
\end{aligned} \tag{6.9}
$$

where $\tilde{x} = x_d - x$ represents the end-effector pose derivation, $M_d, K_d, D_d \succ 0$ are the equivalent mass, spring and damper in Cartesian space. J is the Jacobian matrix[2], h_{ext} are the external wrenches (force and torque) acting on the robot end-effector.

Substituting Eq. (6.9) into tobot dynamic equation Eq. (6.1), it results the following end-effector error dynamic:

$$
M_d\ddot{\tilde{x}} + D_d\dot{\tilde{x}} + K_d\tilde{x} = h_{ext}, \tag{6.10}
$$

The above equation describes a fully decoupled impedance model in Cartesian Space. However, the implementation complexity of such an ideal impedance behavior is high. First of all, measurement of external wrenches is necessary. Secondly, it requires the online computation of the inverse and derivative of the Jacobian matrix. Sometimes an analytical solution even does not exist, so that a numeric solver is needed. Moreover, the robot pose should be kept away from a singular configuration. Otherwise, the inverse of Jacobian could be extremely large. Another prerequisite is that the number of robot DOFs must not be smaller than the size of the impedance matrices.

Similar to joint impedance control, there exists several simplifications on the standard Cartesian impedance control law, which are briefly summarized in Table 6.1.

---

[2]Without specification, all the Jacobian matrices in this thesis are analytical Jacobian.

Control law	Resulted dynamic	Remarks
$\tau = M(q)J^{-1}(q)M_d^{-1}\left(M_d\ddot{x}_d + K\tilde{x} + D\dot{\tilde{x}} - M_d\dot{J}(\dot{q},q)\dot{q}\right)$ $+ g(q) + C(q,\dot{q})\dot{q}$ (6.11)	$M_d\ddot{\tilde{x}} + D\dot{\tilde{x}} + K\tilde{x} =$ $M_dJ(q)M^{-1}(q)J^T(q)h_{ext}$ (6.12)	Coupling effect in $J(q)M^{-1}(q)J^T(q)$.
Same equation above with $M_d = J^{-T}(q)M(q)J^{-1}(q)$ (6.13)	$J^{-T}(q)M(q)J^{-1}(q)\ddot{\tilde{x}}$ $+ K_D\dot{\tilde{x}} + K_P\tilde{x} = h_{ext}$ (6.14)	No inertial shaping, coupling effect in acceleration term, ignorable for small $\ddot{\tilde{x}}$.
$\tau = J^T(q)(K\tilde{x} + D\dot{x}) + g(q) + C(q,\dot{q})\dot{q}$ (6.15)	$-J^{-T}(q)M(q)\ddot{\tilde{q}}$ $+ D_d\dot{\tilde{x}} + K_d\tilde{x} = h_{ext}$ (6.16)	PD control with gravity compensation, more compliance effect, as compliance control for slow motion.

Table 6.1: A briefly summary of several other commonly used Cartesian impedance control laws

Implementation issues

This thesis uses impedance control not only as a motion-level controller but also an equivalent model to describe robot end-effector for further analysis. Hence, it is important to check if the robot demonstrates the desired impedance behavior. Several tests were performed on a *Franka Emika* robot.

At first, the standard impedance control law in Eq. (6.9) was implemented. All the required model parameters are available in the *Franka Control Interface*[3]. The external torques and forces are estimated based on the integrated joint torque sensor and robot dynamic model. However, after several tests, it is concluded that the ideally decoupled impedance control cannot be achieved on this robot platform. The problem mainly lies in the force/torque feedback. As described in the robot documentation, "the external forces and torques are only estimated values that could be inaccurate depending on the robot configuration". Inaccurate force feedback will generate wrong control torque input and cause unexpected or even unstable robot motion. Therefore, it is suggested not to use the decoupled impedance control law in Eq. (6.9) with low-quality force measurements.

Next, the other three control laws in Table 6.1 were tested. The implementation was successful. However, the robot end-effector cannot precisely reproduce the equivalent impedance behavior since coupling effects exist in the resulted dynamics. Their influence depends on joint configurations.

The tests made in this work show that the actual end-effector impedance behavior differs from the desired one during implementation. It is mainly due to the non-linear coupling effect between different DOFs and un-modelled dynamics, e.g. friction. For applications that require precise force/interaction control, an online correction of the impedance behavior is needed. However, to the author's best knowledge, there is little discussion about how to correct the apparent impedance behavior during execution of an impedance control law. One related work [57] mentioned this problem and proposed a method to make the apparent inertia matrix as diagonal as possible (for a decoupled dynamic) through redundancy resolution. Another potential solution would be firstly determining the actual impedance parameter via sensor feedback and online identification, then adjusting the designed impedance matrices (including their non-diagonal elements) to reduce the error between actual and desired impedance behaviors.

---

[3]https://frankaemika.github.io/docs/ Last visit: 15.02.2022

6.3 Analysis of physical human-robot collaboration

As introduced in the last section, impedance control aims to regulate the individual dynamic when interacting with the environment. The main focus of this section is the group behavior. A preliminary analysis of a typical physical human-robot task, namely collaborative transportation of a rigid object, is presented. From the physical point of view, the main characteristic of such collaboration tasks is that each participant is coupled through the manipulated object's dynamic. The coupling effects are described through kinematic constraints and wrench distributions, which dramatically influence task execution and coordination. To further simplify the analysis, each agent (including the human) is modeled as mechanical impedance. Then the collaboration task is formulated as an optimization problem, in which each agent should contribute a wrench (generated by the virtual mechanical impedance) to minimize a task-specified cost function.

6.3.1 System model

Object dynamic

The object's dynamic equation in Cartesian space can be derived from the Lagrange formulation. The total kinetic and potential energy of the object is determined as follows [133]:

$$\mathrm{K} = \frac{1}{2} \left(\dot{\boldsymbol{x}}_o^0 \right)^T \boldsymbol{M}_o \left(\boldsymbol{x}_o^0 \right) \dot{\boldsymbol{x}}_o^0, \tag{6.17}$$

$$\mathrm{P} = -m_o \left(\boldsymbol{g}^0 \right)^T \boldsymbol{p}_o^0. \tag{6.18}$$

K and P represent the kinetic and potential energy respectively, $\boldsymbol{x}_o^0 = (\boldsymbol{p}_o^0, \boldsymbol{\omega}_o^0)^T \in \mathbb{R}^{6 \times 1}$ represents the position and orientation of the object in respect to the world coordinate system, $\boldsymbol{M}_o = \mathrm{diag} \left(m_o \boldsymbol{I}^{3 \times 3}, \mathbf{I}_o \right)$ is the object inertia matrix, in which m_o represent the object's mass, $\mathbf{I}_o \in \mathbb{R}^{3 \times 3}$ is the position-dependent inertial tensor. $\boldsymbol{g}^0 = (0, 0, -g)^T$ is the gravity vector. $\boldsymbol{p}_o^0 \in \mathbb{R}^{3 \times 1}$ describes the position vector of the object's center of mass in the world coordinate system. A graphical illustration is shown in Fig. 6.3.

Taking the derivatives of both kinetic and potential energy, the object's dynamic equation is formalized as:

$$\boldsymbol{M}_o \left(\boldsymbol{x}_o^0 \right) \ddot{\boldsymbol{x}}_o^0 + \boldsymbol{C}_o \left(\boldsymbol{x}_o^0, \dot{\boldsymbol{x}}_o^0 \right) \dot{\boldsymbol{x}}_o^0 + \boldsymbol{h}_g = \boldsymbol{h}_o. \tag{6.19}$$

where

$$\boldsymbol{C}_o \left(\boldsymbol{x}_o^0, \dot{\boldsymbol{x}}_o^0 \right) = \begin{pmatrix} \mathbf{0}^{3 \times 3} & \mathbf{0}^{3 \times 3} \\ \mathbf{0}^{3 \times 3} & \boldsymbol{\omega}_o^0 \times \mathbf{I}_o \end{pmatrix}, \quad \boldsymbol{h}_g = \begin{pmatrix} -m_o \boldsymbol{g}^0 \\ \mathbf{0}^{3 \times 3} \end{pmatrix} \tag{6.20}$$

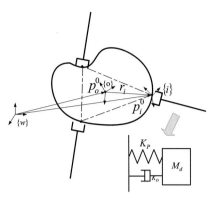

Figure 6.3: Joint transporting a rigid object: sketch of the system

represent the Coriolis and gravity terms. $\boldsymbol{\omega}_o \in \mathbb{R}^3$ describes the object's angular velocity in the world coordinate system $\boldsymbol{h}_o \in \mathbb{R}^{6 \times 1}$ is the total effective wrench acted on the object.

Wrench distribution

The effective wrench $\boldsymbol{h}_o$ generates the object's motion and is produced by all the participants. Assume that each agent i is equipped with a force/torque sensor, and the measurement is performed its own coordinate system, denoted as $\boldsymbol{h}_i \in \mathbb{R}^{6 \times 1}$. Then the following relationship holds:

$$h_i^0 = T_i^0 h_i, \quad T_i^0 = \begin{pmatrix} R_i^0 & 0 \\ 0 & R_i^0 \end{pmatrix}, \tag{6.21}$$

where $\boldsymbol{h}_i^0$ is the same wrench generated by agent i but represented in the world coordinate system. $\boldsymbol{R}_i^0 \in \mathbb{R}^{3 \times 3}$ is the rotation matrix of coordinate system i in respect to the word coordinate system.

Furthermore, assuming that the grasping of the object by each agent i is rigid (no relative motion between the gripper and the grasping point), the distribution of all the wrenches generated by totally N agents on the rigid object is determined by:

$$h_o = G h_\Sigma^0, \quad h_\Sigma^0 = \begin{pmatrix} h_1^0 & \dots & h_N^0 \end{pmatrix}^T. \tag{6.22}$$

G is named as grasp matrix. According to the virtual link model proposed in [134], it is formulated as follows:

$$G = \begin{pmatrix} I & 0 & I & 0 & ... & I & 0 \\ S(r_1) & I & S(r_2) & I & ... & S(r_N) & I \end{pmatrix} \in \mathbb{R}^{6 \times 6N}, \quad (6.23)$$

where $S(\cdot) \in \mathbb{R}^{3 \times 3}$ is the skew-symmetric matrix, performing the cross-product operation. r_i describes the relative position vector between the object's center of mass and the grasping point of agent i in the object's coordinate system (see Fig. 6.3). Note that the grasp matrix remains constant as long as the grasping points are fixed.

It can be seen that due to several zeros in the grasp matrix G, not all components in h_Σ^0 are transferred into object's motion. The rest part generates "internal wrenches", which are determined by the null-space projection of the grasp matrix, which yields:

$$h_{\text{int}}^0 = \left(I - G^\dagger G \right) h_\Sigma^0, \quad (6.24)$$

where $G^\dagger$ represents the pseudo-inverse of G. The internal wrenches have no contribution to the motion but can cause deformation or even damage the object. Hence, many control designs aim to minimize their values. Note that internal wrenches cannot be directly measured. When the forces/torques generated by each agent are known, they can be calculated using the equation above.

Kinematic constraints

In the case of rigid grasping of a rigid object, each agent cannot move arbitrarily even within their workspaces. Their motions are coupled through the kinematic constraints. Firstly, the following relationship holds for the positions :

$$p_i^0 = p_o^0 + R_o^0 r_i \quad (6.25)$$

where R_o^0 is the rotation matrix between the object and world frame.

Taking the first and second derivation of the above equation, the translational velocity and acceleration constraints are determined as follows:

$$\dot{p}_i^0 = \dot{p}_o^0 + \omega_o^0 \times r_i, \quad (6.26)$$
$$\ddot{p}_i^0 = \ddot{p}_o^0 + \dot{\omega}_o^0 \times r_i + \omega_o^0 \times \left(\omega_o^0 \times r_i \right), \quad (6.27)$$

where ω_o^0 represents the object's angular velocity.

The rotational motions of each agent and the object in respect to the world coordinate system should satisfy:

$$\omega_i^0 = \omega_o^0, \tag{6.28}$$

$$\dot{\omega}_i^0 = \dot{\omega}_o^0. \tag{6.29}$$

One practical application of the kinematic constraints is to estimate motions with only partially available information on the system. For instance, if the effector's motion of each agent is tracked well by a marker-based optical motion capture system in a global coordinate frame, then the object's motion can be determined based on the kinematic constraints without installing extra markers. Another typical setup in practice is an entirely mobile measurement system, in which only "local" information according to each agent's coordinate system is available. Still, one agent can combine the kinematic constraints with its measurements to estimate the motions of the others. More details can be found in one related work [135].

Agent dynamic

Within the concept of this work, each robot in the team is controlled through an impedance controller, in which its end-effector's tracking error dynamic is equivalently interpreted through the mechanical impedance (mass, spring, and damper). Assuming that each robot i aims to following a reference trajectory defined by $\ddot{x}_{i,d}, \dot{x}_{i,d}, x_{i,d}$, it yields:

$$M_{i,d}\ddot{\tilde{x}}_i + D_{i,d}\dot{\tilde{x}}_i + K_{i,d}\tilde{x}_i = h_i - h_{ext,i}, \tag{6.30}$$

where $M_{i,d}, D_{i,d}$ and $K_{i,d}$ are the desired mass, damping and stiffness, $\tilde{x}_i = x_{i,d} - x_i$ represents the end-effector motion error of the robot control, and $h_{ext,i}$ represents the external wrench acted on robot's end-effector during the collaboration.

One pioneering work [136] provides the evidence that when interacting with the physical environment, the human motor control system generates also acts like mechanical impedance, especially the spring-like property of muscles. Hence, the human arm dynamic can be modeled as a set of springs, which generate static forces to compensate for the position tracking error. A further extension with a damping term is also a popular design in literature and has been validated through experiments [137, 138]. In this work, the human end-effector dynamic is modeled as a spring-damper system with velocity feedback (PD-type controller), so that:

$$D_H\dot{\tilde{x}}_H + K_H\tilde{x}_H = h_H - h_{ext,H}, \tag{6.31}$$

where K_H and D_H are the human impedance parameters, and $\tilde{x}_H$ represents the human

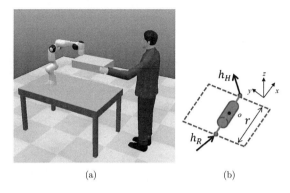

(a)　　　　　　　　　　　　　　　(b)

Figure 6.4: Graphical illustration of the physical human-robot collaboration scenario considered in this work: (a) overview, (b) grasping configuration under the virtual link model.

position control error, $\boldsymbol{h}_H$ is the wrench generated by human and $\boldsymbol{h}_{ext,H}$ represents the external wrench acted on human hand.

Note that the impedance model can also be applied for the description of human joint dynamic, as long as the measurement of joint movements is available [139]. Moreover, it should be pointed out that most literature concludes that the human impedance parameters vary during the task execution. Hence, learning/identifying these parameters is critical for online use.

Based on the analysis above, the PHRC scenario can be physically interpreted as shared impedance control, in which both the human and the robot are modeled as mechanical impedance and generate force/torque to influence the object's dynamic. Under this concept, role allocation can be achieved by varying their impedance parameters. For example, the human would dominate the control by increasing its impedance to generate large forces. In short, the dynamic role allocation in PHRC could be formulated as an adaptive impedance control problem.

Case study: dyadic human-robot team

This work considers a physical collaboration task with one human and one robot as a team (see Fig. 6.4(a)). The sensor system is only mounted on the robot so that all the measurements are recorded in respect to the robot coordinate system (see Fig. 6.4(b)). The distance between the two grasping points is denoted as r. A further assumption is that the object's center of mass locates on the middle point between the two grasping points. Under this concept, the human grasping point can be regarded as a extension of the robot's end-effector along its x-axis. Denoting the robot end-effector pose as $\boldsymbol{x}_R = (\boldsymbol{p}_R, \boldsymbol{\theta}_R)^T \in \mathbb{R}^{6\times1}$, the human hand pose $\boldsymbol{x}_H = (\boldsymbol{p}_H, \boldsymbol{\theta}_H)^T$ can be determined with the kinematic constraints, namely:

$$\boldsymbol{p}_H = \boldsymbol{p}_R + \boldsymbol{R}\left(\boldsymbol{\theta}_R\right) \begin{pmatrix} r \\ 0 \\ 0 \end{pmatrix}, \quad \boldsymbol{\theta}_H = \boldsymbol{\theta}_R. \tag{6.32}$$

The grasping matrix is determined as follows:

$$\boldsymbol{G} = \begin{pmatrix} \boldsymbol{G}_H & \boldsymbol{G}_R \end{pmatrix}, \quad \boldsymbol{G}_H = \begin{pmatrix} \boldsymbol{I} & \boldsymbol{0} \\ \boldsymbol{S}_H & \boldsymbol{I} \end{pmatrix}, \boldsymbol{G}_R = \begin{pmatrix} \boldsymbol{I} & \boldsymbol{0} \\ \boldsymbol{S}_R & \boldsymbol{I} \end{pmatrix},$$

$$\boldsymbol{S}_H = \begin{pmatrix} 0 & 0 & 0 \\ 0 & 0 & \frac{r}{2} \\ 0 & -\frac{r}{2} & 0 \end{pmatrix}, \boldsymbol{S}_R = \begin{pmatrix} 0 & 0 & 0 \\ 0 & 0 & -\frac{r}{2} \\ 0 & \frac{r}{2} & 0 \end{pmatrix}. \tag{6.33}$$

In a quasi-static case where the object moves slowly, i.e. $\ddot{\boldsymbol{x}}_o, \dot{\boldsymbol{x}}_o \approx \boldsymbol{0}$, the wrench generated by human can be estimated based on the object dynamic (Eq. (6.19)):

$$\hat{\boldsymbol{h}}_H = \boldsymbol{G}_H^{-1}\left(\boldsymbol{h}_g - \boldsymbol{G}_R\boldsymbol{h}_R\right). \tag{6.34}$$

6.3.2 Problem formulation

The collaboration task for the human-robot team is formulated as an optimal control problem, in which both human and robot should generate optimal wrenches to minimize the following cost function:

$$\min_{\boldsymbol{h}_\Sigma^0} \int_0^\infty \left(\mathbf{X}^T \mathbf{Q} \mathbf{X} + \left(\boldsymbol{h}_\Sigma^0\right)^T \mathbf{R} \boldsymbol{h}_\Sigma^0\right) \mathrm{d}t,$$

s.t. object dynamic in Eq. (6.19), $\tag{6.35}$

where $\mathbf{X} = \left(\tilde{\boldsymbol{x}}_o^0, \dot{\boldsymbol{x}}_o^0\right)^T$ is the object's state variable, $\tilde{\boldsymbol{x}}_o^0 = \boldsymbol{x}_{o,d}^0 - \boldsymbol{x}_o^0$ represents the position control error. $\mathbf{Q}, \mathbf{R} \succ 0$ are diagonal positive definite weighting matrices.

Eq. (6.35) can be regarded as a "global" optimization problem, which aims to minimize the object's position tracking error, the kinetic energy, and the total effort injected into the system. However, solving the global optimal control problem in a centralized way is difficult in practice. The system model is nonlinear and requires full knowledge of the object's geometric (center of mass, grasping positions) and dynamic parameters (mass, inertial). Even though a central controller could be designed to compute the optimal wrench that each agent in the team should generate, the human as an "actuator" cannot achieve precise force control. Moreover, the problem formulation does not include any description of interactive behaviors between agents.

Previous research [129] suggests using differential game theory to describe interactions of two agents when performing a joint motor task. It is assumed that each agent in the game has complete knowledge of the system dynamic and the others' control input. Considering a dyadic human-robot team, the robot needs to solve the following optimization problem:

$$\min_{\boldsymbol{h}_R} \int_0^\infty \left(\mathbf{X}^T \mathbf{Q} \mathbf{X} + \boldsymbol{h}_R^T \mathbf{R}_R \boldsymbol{h}_R + \boldsymbol{h}_H^T \mathbf{R}_H \boldsymbol{h}_H \right) \mathrm{d}t,$$

s.t. object dynamic in Eq. (6.19) $\hspace{3cm}$ (6.36)

where $\boldsymbol{h}_R$ and $\boldsymbol{h}_H$ represents the control input of the robot and the human respectively. In this work they are evaluated by force measurements. The weighting matrices $\mathbf{Q}, \mathbf{R}_R$ and $\mathbf{R}_H$ can be chosen differently by each agent. The first two terms are the same as the global task, namely minimizing the object's pose error and total energy consumption. The last term describes a "collaborative" behavior since the robot also tries to minimize the control effort of the human [129].

In the game theory, the solution of a differential game can be defined by Nash equilibrium, Stackelberg equilibrium, or Pareto optimum [4]. Authors in [140] argue that Nash equilibrium is more suitable to describe a collaborative physical human-robot interaction, in which the roles of human and robot are equally distributed. At Nash equilibrium, each agent cannot achieve better performance by changing its control policy (the impedance control gains). If the human-robot team follows a leader-follower relationship, then Stackelberg equilibrium would be a better description since it provides that the follower should optimize its cost function sequentially after the leader. Pareto optimum provides that each agent optimizes its cost function without harming the others. It requires more coordination and information exchange between the agents, which is difficult to achieve in practice, especially for a hybrid human-robot team.

Analytically the Nach equilibrium can be determined by solving the Hamilton-Jacobian-Bellman (HJB) equation [141]. Denoting the robot's cost function as :

$$r_R = \mathbf{X}^T \mathbf{Q} \mathbf{X} + \boldsymbol{h}_R^T \mathbf{R}_R \boldsymbol{h}_R + \boldsymbol{h}_H^T \mathbf{R}_H \boldsymbol{h}_H, \hspace{2cm} (6.37)$$

then the HJB equation is formulated as:

$$H_R\left(\mathbf{X}, \nabla V_R, \boldsymbol{h}_R, \boldsymbol{h}_H\right) = r_R\left(\mathbf{X}, \boldsymbol{h}_R, \boldsymbol{h}_H\right) + \left(\nabla V_R\right)^T \boldsymbol{f}\left(\mathbf{X}, \boldsymbol{h}_R, \boldsymbol{h}_H\right) = 0, \qquad (6.38)$$

where H_R is called Hamilton function, $\boldsymbol{f}$ is the state-space representation of the system dynamic, namely:

$$\dot{\mathbf{X}} = \boldsymbol{f}\left(\mathbf{X}, \boldsymbol{h}_R, \boldsymbol{h}_H\right)$$

$$= \begin{pmatrix} \mathbf{0} & -\boldsymbol{I} \\ \mathbf{0} & \left(\boldsymbol{M}_o\right)^{-1}\boldsymbol{C}_o \end{pmatrix} \mathbf{X} + \left(\boldsymbol{M}_o\right)^{-1}\left(\boldsymbol{G}_H\boldsymbol{h}_H + \boldsymbol{G}_R\boldsymbol{h}_R\right). \qquad (6.39)$$

Under a further assumption that the robot's control input $\boldsymbol{h}_R$ is unconstrained, the solution of the optimization problem is obtained by solving:

$$\frac{\partial H_R}{\partial \boldsymbol{h}_R} = \mathbf{0}, \qquad (6.40)$$

which results in the following analytical solution:

$$\boldsymbol{h}_R^\star = -\frac{1}{2}\left(\mathbf{R}_R\right)^{-1}\boldsymbol{G}_R^T\nabla V_R^\star. \qquad (6.41)$$

Different solution approaches are possible to determine the optimal control input $\boldsymbol{h}_R^\star$. For instance, authors in [142] reformulate the problem in the form of linear quadratic control (LQR) with several simplifications and compute the optimal state-feedback control gain by solving the Riccati equation. Authors in [143] argue that when the system is nonlinear, a direct solution of the Riccati equation is difficult or sometimes even impossible. Hence, they present a policy iteration algorithm that iteratively solves the HJB equation. Still, both approaches require full knowledge of the system dynamic, which is not always available in practice and makes the solution inflexible. Moreover, the error in models also influences the solution of the optimization problem. To deal with these issues, an online method based on reinforcement learning without an explicit knowledge on the system dynamic will be presented in the next chapter.

6.4 Summary

This chapter introduced several fundamentals of physical human-robot collaboration, including system modeling, problem formulation, and control strategy. One of the most common benchmarking applications, namely collaborative transportation of an object, was studied. During execution of the task, each agent contributes forces/torques to generate the object's motion and keep a task-dependent group formation. By analyzing the object's dynamic model, it is concluded that kinematic and dynamic constraints dramatically influence the coordination between agents. On the one hand, the constraints limit the movement in some DOFs, making agents' motion less uncertain and easier to predict. On the other hand, proper coordination and compliant control strategies are required to avoid large internal wrenches generated by motion conflicts.

Impedance control provides a way of achieving compliant behavior. The basic concept is to modulate the robot joint or end-effector dynamics as a mass-spring-damper system. It should be pointed out that an ideal configuration-independent decoupled impedance behavior is difficult to realize in practice since it requires high-quality force/torque measurements and a precise robot dynamic model. Therefore, there exist several alternative impedance control laws, which come up with a compromise between accuracy and labor expenses. Online regulation of the apparent impedance behavior would be interesting to study in future work.

Previous research shows that mechanical impedance can also describe human motor control dynamics during physical interaction, which significantly simplifies the modeling and analysis of the whole system. Moreover, recalling the DMP formulation in Chapter 4, it can be seen that the human impedance model is equivalent to the linear goal-attraction term in DMP. The nonlinear shape attraction term is replaced by the real existing interaction force. This correlation shows that mechanical impedance is a main determinant of interactions and can be applied in a wide range of collaboration tasks (human-human or human-robot, with or without physical contact, etc.)

After analyzing the system dynamic from both the physical and compliance control point of view, the human-robot collaboration task is formulated as a differential game, in which both human and robot aim to find an optimal strategy to minimize an objective function. Game theory provides a mathematical framework that can represent different types of interactions, e.g., cooperative, non-cooperative, leader-follower, equal role-assignment, etc. The solution is usually given by optimal control theory, which yields an optimal feedback control gain (here: impedance parameters). A direct approach for solving the optimization problem is not always possible. In the next chapter, a learning-based method based on an adaptive impedance control framework will be presented.

7 Adaptive Impedance Control based on Reinforcement Learning

7.1 Introduction

As discussed in the previous chapter, impedance control plays a dominant role in describing and executing physical human-robot collaboration (PHRC). In such tasks, when and how the interactions occur might vary significantly during the execution. Hence, conventional impedance control with constant parameters is insufficient.

Previous research aims to find a proper adaptive mechanism for the impedance parameters according to different input perceptions. In [57] the damping factor is tuned based on the human guiding velocity, i.e., high damping is provided when the human performs fine movements at low velocity, while low damping is used for large human movements at high velocity. In [58] the impedance parameters are adjusted by taking feedback of contact force, following a "large force $\to$ low damping" rule. [59] considers a collaborative sawing task, in which the robot adapts its stiffness based on the EMG signal feedback from sensors mounted on the human arm to generate varying supportive forces. Moreover, optical information is used to estimate human arm pose and refine the robot stiffness distribution in different directions. In [144] the authors first study the human motor control scheme and then transfer it into the robot so that a human-like adaption on the impedance parameters is achieved.

Note that all these works have developed an adaptive impedance control that intuitively reacts to the current sensory feedback. However, there is still a lack of consideration of task accomplishment and long-term performance. The previous chapter introduced the concept of formulating the collaboration task as an optimization problem, in which different task-relied and human-relied features are combined in the cost function. Thereby, this chapter aims to find the optimal control policy, or in other words, the optimal impedance parameters, by solving the problem online.

Conventional optimal control is usually model-based and non-adaptive. It is not sufficient for PHRC since there exist high uncertainties in human control goals and control mechanisms. Machine learning techniques, especially reinforcement learning (RL), provide a promising solution. As reviewed in [55], a combination of RL and impedance control has been studied in various research works. In [145], a path-integral-based RL approach is developed to optimize joint impedance parameters. The key concept is to use dynamic movement primitives for parameterizing and learning optimal joint stiffnesses. In [140] the authors proposed an RL-based adaptive control in a PHRC task. An actor-critic structure with two neural networks is designed to estimate human control policy and simultaneously update the robot control. In [146], a cascaded RL-based impedance control loop is developed. A task-specific outer-loop aims to learn an optimal impedance model, while a robot-specific inner-loop controller aims to achieve the prescribed impedance behavior. [147], presents the design of a fuzzy Q-Learning algorithm to regulate the desired damping vector by minimizing the jerk of the robot motion. Note that all the above-mentioned related work does not require an accurate model of the environment dynamics.

This chapter presents a novel RL-based approach for adaptive impedance control. Firstly, Q-learning with quadratic value function is used to determine the optimal control policy (impedance parameters) online based on the motion data measurements. Secondly, the learned impedance parameters are further adjusted by taking feedback of the interaction force, which is regarded as the degree of human's acceptance on the robot's control action.

The rest of this chapter is organized as follows. Section 7.2 briefly reviewed the basic concept of RL and the most commonly studied approaches to solving an RL problem. Section 7.3 presents a Q-learning-based adaptive impedance control method, including the mathematical formulation, the physical interpretation, and discussions on stability and constraints handling. The learning algorithm is validated with numerical examples. Section 7.4 summarizes the chapter.

7.2 Basics of reinforcement learning

Reinforcement learning (RL) is a machine learning approach concerned with how an agent learns a behavior by interacting with its environment in order to maximize the total amount of reward it receives over a certain time [148]. Fig. 7.1[1] illustrates the general structure of RL with the most critical elements. At each time step t, the agent receives feedback from its environment, denoted as *state*, as a result from the last *action*. Then an evaluation is performed using a performance index, denoted as *reward*. Afterward,

---

[1]Image source: https://www.slideserve.com/sumana/reinforcement-learning Last view: 20.02.2022

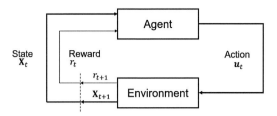

Figure 7.1: Principle of reinforcement learning

the agent determines a *policy* in order to maximize the cumulative rewards in the future, represented by a *value function*. Lastly, following the policy, the agent generates a new action to interact with the environment and receives new feedback at time step $t + 1$.

RL has been widely studied by different societies such as computer science, machine learning, and control engineering in the last decades [141, 148, 149]. This thesis discusses RL from the approximate dynamic programming (ADP) point of view.

Note that in this thesis, the performance is evaluated by a cost function (negative reward). Considering the following infinite-horizon optimization problem in discrete form:

$$V^* (\mathbf{X}_t) = \min_{\boldsymbol{u}} \sum_{k=0}^{\infty} \gamma^k r (\mathbf{X}_{t+k}, \boldsymbol{u}_{t+k}), \tag{7.1}$$

$$\text{s.t. } \mathbf{X}_{t+k+1} = \boldsymbol{F} (\mathbf{X}_{t+k}) + \boldsymbol{B} (\mathbf{X}_{t+k}) \boldsymbol{u}_{t+k}, \tag{7.2}$$

where V describes the sum of future cost starting from t. r denotes the stage cost which is determined by the current state and control input. $\gamma \in (0, 1]$ is a discount factor that reduces the weight of costs in the future. Eq. (7.2) describes the system dynamic. Note that the problem formulation is in contrast to standard RL, i.e. instead of maximizing a value function as a sum of stage rewards, the purpose turns to minimize a cost function (negative rewards).

Separating the first sum term, the value function V can be reformulated as:

$$V (\mathbf{X}_t) = r (\mathbf{X}_t, \boldsymbol{u}_t) + \gamma \sum_{k=0}^{\infty} \gamma^k r (\mathbf{X}_{t+k+1}, \boldsymbol{u}_{t+k+1}). \tag{7.3}$$

Furthermore, assuming that the control input $\boldsymbol{u}$ is determined by a state feedback control policy, denoted as $\boldsymbol{\mu}$, so that:

$$\boldsymbol{u}_t = \boldsymbol{\mu} (\mathbf{X}_t). \tag{7.4}$$

Substituting the above equation into Eq. (7.3), it yields the following difference equation of the value function:

$$V\left(\mathbf{X}_t\right) = r\left(\mathbf{X}_t, \boldsymbol{\mu}\left(\mathbf{X}_t\right)\right) + \gamma V\left(\mathbf{X}_{t+1}\right) \tag{7.5}$$

According to Bellman's *principle of optimality*, the optimal value function V^* should satisfy:

$$V^*\left(\mathbf{X}_t\right) = \min_{\boldsymbol{\mu}} \left(r\left(\mathbf{X}_t, \boldsymbol{\mu}\left(\mathbf{X}_t\right)\right)\right) + \gamma V^*\left(\mathbf{X}_{t+1}\right). \tag{7.6}$$

The meaning of the above equation is that "an optimal policy has the property that no matter what the previous control actions have been, the remaining controls constitute an optimal policy with regard to the state resulting from those previous controls".[150] Note that this equation is also denoted as Hamilton-Jacobian-Bellman (HJB) equation, and can be regarded as the discrete form of Eq. (6.38).

Now it comes to the same problem that remained in the previous chapter: how can the HJB equation be solved to obtain the optimal control policy $\boldsymbol{\mu}^*$ through RL?

There is a large family of algorithms for solving an RL problem. As reviewed in e.g. [149, 151, 152], they can be roughly classified in three approaches: (1) model-based, (2) value-function-based, (3) policy-based. Fig. 7.2 provides a graphical illustration of their principles.

The model-based approach is close to the concept of optimal control (sometimes even identical), e.g., dynamic programming (DP), model predictive control (MPC) [153]. The prerequisite is that the system model in Eq. 7.2 is known. If the system is linear time-invariant, then the optimal policy can be analytically determined by solving the *algebraic Riccati equation* (ARE). Otherwise, there might exist no global optimal solution, but the model still enables prediction of the future states and thereby updates the value function or the policy. Model-based approaches usually require a small number of interactions between agent and environment, and converges fast to optimal solutions. However, the performance strongly relies on the model's accuracy.

The value-function-based approach aims to learn the value function through function approximators and use it to reconstruct the optimal policy. A variety of methods of value-function-based reinforcement learning algorithms have been developed. In model-free cases, they can be roughly split into Monte Carlo methods and temporal difference (TD) learning. The former estimates the value function through "roll-outs" by executing the current control policy. The latter learns the value function based on temporal error, namely the difference between the old and the new estimations of the value function. As long as the optimal value function V^* is learned, the globally optimal policy follows simply by calculating the gradient of V^*. However, a well-known disadvantage of the value-function-based approach is the curse of dimensionality. Hence, this approach is usually suitable for low-dimensional systems.

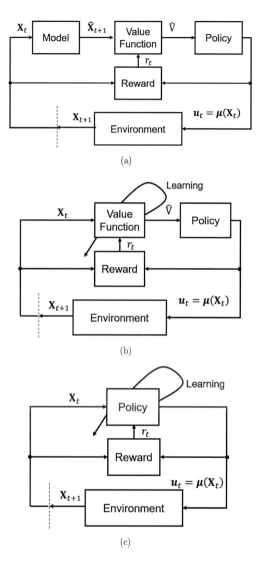

Figure 7.2: Graphical illustration of the principles of three RL approaches: (a) model-based, (b) value-function-based, (c) policy based.

As the name implies, the policy-based approach directly calculates the optimal policy by modifying its parameters. The key step is the computation of the policy update rules which can be solved by gradient-method, stochastic optimization, information theory, etc. Compared to the value-function-based approach, the main advantage of the policy-based approach is its scalability in high-dimensional systems. However, there is no guarantee that the results converge to global optimality. Moreover, during the policy-search process, the agent needs to interact with the environment frequently and might even result in dangerous states which are critical in the context of human-robot collaboration.

In this work, a value-function-based approach with TD learning is implemented for the following reasons: (1) easy implementation due to small dimensions, (2) clear interpretation with the guarantee of convergence, (3) possible combination with Cartesian impedance control. This will be discussed in more detail in the next section.

7.3 Q-Learning-based adaptive impedance control

7.3.1 Unconstrained Learning

Problem formulation

This section still focuses on the task that a human and a robot collaboratively transport a rigid object. To simplify the analysis, only translational motions are considered. Then both the object and agent dynamics in Eq. (6.19) and Eq. (6.30) become linear. And the kinematic constraints are simplified as:

$$\begin{aligned}
\boldsymbol{p}_R &= \boldsymbol{p}_o + \text{const.}, \\
\boldsymbol{p}_H &= \boldsymbol{p}_o + \text{const.}, \\
\dot{\boldsymbol{p}}_H &= \dot{\boldsymbol{p}}_R = \dot{\boldsymbol{p}}_o,
\end{aligned} \tag{7.7}$$

where $\boldsymbol{p} \in \mathbb{R}^{3 \times 1}$ represents position, the indexes H, R, o represent "human","robot" and "object" respectively. Thereby, the cost function defined in Eq. (6.37) is rewritten in the following equation:

$$r\left(\mathbf{X}_R, \boldsymbol{u}_R\right) = \mathbf{X}_R^T \mathbf{Q}_R \mathbf{X}_R + \boldsymbol{u}_R^T \mathbf{R}_R \boldsymbol{u}_R, \tag{7.8}$$

where $\mathbf{X}_R = (\tilde{\boldsymbol{p}}_R, \dot{\boldsymbol{p}}_R) \in \mathbb{R}^{6 \times 1}$, $\tilde{\boldsymbol{p}}_R = \boldsymbol{p}_R - \boldsymbol{p}_{R,d}$ represents the position tracking error, $\boldsymbol{u}_R \in \mathbb{R}^{3 \times 1}$ is a "virtual" robot control input in Cartesian space.

Remarks Note that in comparison to (6.37), the above cost function has two main differences. Firstly, all the object states are replaced by the robot states. It would not influence the result of the optimal control policy since the robot position and object position only differ in terms of a constant offset, and their velocities are equal. Secondly, the term relied on the human control input (force) $\boldsymbol{h}_H$ is removed since human force sensing is not available. In this work, human force is indirectly estimated from the external torques measured by robot joint torque sensors and used as a corrector of the learned control policy. Note that it is a suboptimal solution due to hardware limitations. If the human force sensing is available, it is suggested to include the measurements in the robot cost function to avoid estimation errors [140, 146]. At last, it should be mentioned that one recent work [142] proposed a robot controller that is able to estimate human control gain without force sensor. However, the method requires full knowledge of the interaction dynamic, including the inertia and viscosity matrices of the object.

$V(\mathbf{X}_R)$ is written in a continuous-time form as integral of the reward function, namely:

$$V(\mathbf{X}_R) = \int_t^\infty r(\mathbf{X}_R, \boldsymbol{u}_R)\,\mathrm{d}\tau. \tag{7.9}$$

Now, the optimization problem is to find an optimal control input $\boldsymbol{u}_R^*$ so that:

$$\boldsymbol{u}_R^* = \arg\min_{\boldsymbol{u}_R} \int_t^\infty r(\mathbf{X}_R, \boldsymbol{u}_R)\,\mathrm{d}\tau. \tag{7.10}$$

In [141], optimizing the above value function is classified as an integral RL problem. One possible solution approach is to discretize the continuous-time HJB equation:

$$0 = r_R(\mathbf{X}_R, \boldsymbol{u}_R) + (\nabla V_R)^T (\boldsymbol{f}(\mathbf{X}_R, \boldsymbol{u}_R) + \boldsymbol{B}(\mathbf{X}_R)) = r(\mathbf{X}_R, \boldsymbol{u}_R) + \dot{V}^{\boldsymbol{u}_R}. \tag{7.11}$$

Using Euler method, a discrete-time equation is obtained:

$$0 = r(\mathbf{X}_R(k), \boldsymbol{u}_R(k)) + \frac{V(k+1) - V(k)}{\Delta t}$$
$$\Leftrightarrow V(k) = r(k)\Delta t + V(k+1), \tag{7.12}$$

where Δt represents the sampling time. Note that the above equation has the same form as the Bellman equation (Eq. (7.6)). Therefore, all the RL methods developed for discrete-time systems can be applied. However, it is only an approximated approach.

Q-Learning

From now on, an action-value function (Q-function) is used instead of state value function V for further analyses. Learning the Q-function (Q-learning) is classified as an off-policy method, which means that the optimal value function is estimated independently from the current policy [154]. More importantly, as presented in [148], when the model of the environment is unknown, a state-action value function is preferred since it dramatically simplifies the analysis of the algorithm.

The Q-function is defined as follows:

$$
\begin{aligned}
Q\left(\mathbf{X}_R(k), \boldsymbol{u}_R(k)\right) = {} & r\left(\mathbf{X}_R(k), \boldsymbol{u}_R(k)\right) \Delta t \\
& + \gamma Q\left(\mathbf{X}_R(k+1), \boldsymbol{\mu}\left(\mathbf{X}_R(k+1)\right)\right),
\end{aligned}
\tag{7.13}
$$

where $\boldsymbol{\mu}$ represents the control policy. Note that if $\boldsymbol{u}_R(k) = \mu\left(\mathbf{X}_R(k)\right)$, the Q-function is identical to the state value function, namely:

$$
Q\left(\mathbf{X}_R(k), \mu\left(\mathbf{X}_R(k)\right)\right) = V\left(\mathbf{X}_R(k)\right).
\tag{7.14}
$$

Based on Eq. (7.13), the optimal Q-function is defined as:

$$
Q^*\left(\mathbf{X}_R(k), \boldsymbol{u}_k\right) = r\left(\mathbf{X}_R(k), \boldsymbol{u}_k\right) + \gamma V^*\left(\mathbf{X}_R(k+1)\right).
\tag{7.15}
$$

Substituting the above equation into Eq. (7.6), it yields:

$$
V^*\left(\mathbf{X}_R(k)\right) = \min_{\boldsymbol{\mu}} Q^*\left(\mathbf{X}_R(k), \boldsymbol{u}_k\right).
\tag{7.16}
$$

Hence, optimizing the state-value function V is equivalent to optimizing the Q-function.

As discussed in the previous chapter, under the concept of impedance control, the robot end-effector dynamic can be represented by a mass-spring-damper system which results in the following linear discrete-time state-space model:

$$
\mathbf{X}_R(k+1) = \boldsymbol{A}_d \mathbf{X}_R(k) + \boldsymbol{B}_d \boldsymbol{u}_R(k),
\tag{7.17}
$$

where

$$
\boldsymbol{A}_d \approx \boldsymbol{I} + \boldsymbol{A}\Delta t = \boldsymbol{I} + \begin{pmatrix} \mathbf{0} & \boldsymbol{I} \\ -\boldsymbol{M}_d^{-1}\boldsymbol{K}_d & -\boldsymbol{M}_d^{-1}\boldsymbol{D}_d \end{pmatrix} \Delta t,
\tag{7.18}
$$

$$
\boldsymbol{B}_d \approx \left(\boldsymbol{I} + \frac{\Delta t}{2}\boldsymbol{A}\right) \Delta t \begin{pmatrix} \mathbf{0} \\ -\boldsymbol{M}_d^{-1} \end{pmatrix}.
\tag{7.19}
$$

Based on the superposition principle of linear systems, the Q-function can be written as:

$$
Q\left(\mathbf{X}_R(k), \boldsymbol{u}_R(k)\right)
$$

$$
\begin{aligned}
&= \left(\mathbf{X}_R^T(k)\mathbf{Q}_R\mathbf{X}_R(k) + \boldsymbol{u}_R^T(k)\mathbf{R}_R\boldsymbol{u}_R(k) \right) \cdot \Delta t \\
&\quad + \gamma \left(\boldsymbol{A}_d\mathbf{X}_R(k) + \boldsymbol{B}_d\boldsymbol{u}_R(k) \right)^T \boldsymbol{P} \left(\boldsymbol{A}_d\mathbf{X}_R(k) + \boldsymbol{B}_d\boldsymbol{u}_R(k) \right) \\
&= \begin{pmatrix} \mathbf{X}_R(k) \\ \boldsymbol{u}_R(k) \end{pmatrix}^T \underbrace{\begin{pmatrix} \boldsymbol{S}_{xx} & \boldsymbol{S}_{xu} \\ \boldsymbol{S}_{xu}^T & \boldsymbol{S}_{uu} \end{pmatrix}}_{\boldsymbol{S}} \begin{pmatrix} \mathbf{X}_R(k) \\ \boldsymbol{u}_R(k) \end{pmatrix},
\end{aligned} \tag{7.20}
$$

where $\boldsymbol{P} \in \mathbb{R}^{6 \times 6}$ and $\boldsymbol{S} \in \mathbb{R}^{9 \times 9}$ are parameter matrices. The above derivations show that if the system model is linear, the Q function results in a quadratic form.

Next, the value-function-approximation (VPA) technique is used to parameterize and learn the Q-function. A quadratic Q-function can be rewritten as:

$$
Q\left(\mathbf{X}_R(k), \boldsymbol{u}_R(k)\right) = \boldsymbol{S}^T \boldsymbol{\varphi}\left(\mathbf{X}_R(k), \boldsymbol{u}_R(k)\right), \tag{7.21}
$$

where

$$
\boldsymbol{\varphi}\left(\mathbf{X}_R(k), \boldsymbol{u}_R(k)\right) = \begin{pmatrix} \mathbf{X}_R(k) \otimes \mathbf{X}_R(k) \\ 2\mathbf{X}_R(k) \otimes \boldsymbol{u}_R(k) \\ \boldsymbol{u}_R(k) \otimes \boldsymbol{u}_R(k) \end{pmatrix} \tag{7.22}
$$

$$
\boldsymbol{S} = \begin{pmatrix} \operatorname{vec}\left(\boldsymbol{S}_{xx}\right) & \operatorname{vec}\left(\boldsymbol{S}_{xu}\right) & \operatorname{vec}\left(\boldsymbol{S}_{uu}\right) \end{pmatrix}^T. \tag{7.23}
$$

The symbol $\otimes$ represents Kronecker product. $\boldsymbol{S}$ stackes all the elements in the $\boldsymbol{S}$ matrix in one vector.

Minimum parameter set Note that matrix $\boldsymbol{S}$ is symmetric, there exist several identical elements. In general cases, if $\mathbf{X} \in \mathbb{R}^{n \times 1}$ and $\boldsymbol{u} \in \mathbb{R}^{m \times 1}$, the minimum parameter set should have a size of $(n + m)(n + m + 1)/2$ [155]. Hence, the parameter vector $\boldsymbol{S}$ and the feature function $\boldsymbol{\varphi}$ has a length of $(6 + 3)(6 + 3 + 1)/2 = 45$.

The optimal Q-function can be learned through temporal difference (TD) methods. A TD error denoted the difference between the old and new estimation of the Q-function in each step, which is determined as follows:

$$
\begin{aligned}
e(k) = &- Q\left(\mathbf{X}_R(k), \boldsymbol{u}_R(k)\right) + r\left(\mathbf{X}_R(k), \boldsymbol{u}_R(k)\right) \\
&+ \gamma Q\left(\mathbf{X}_R(k+1), \boldsymbol{\mu}(\mathbf{X}_R(k+1))\right).
\end{aligned} \tag{7.24}
$$

Since Q is parameterized by $\boldsymbol{S}$, the above equation is equivalent to:

$$
e(k) = - \boldsymbol{S}^T \boldsymbol{\varphi}\left(\mathbf{X}_R(k), \boldsymbol{u}_R(k)\right) + r\left(\mathbf{X}_R(k), \boldsymbol{u}_R(k)\right)
$$

$$+ \gamma \boldsymbol{S}^T \boldsymbol{\varphi} \left(\mathbf{X}_R(k+1), \boldsymbol{u}_R(k+1) \right). \tag{7.25}$$

Then an optimization problem of minimizing the TD error is defined as:

$$\min_{\boldsymbol{S}} \frac{1}{2} e^2(k), \tag{7.26}$$

which can be solved via gradient decent method. The gradient of the cost function in respect to parameter set $\boldsymbol{S}$ is calculated as:

$$\begin{aligned} \frac{\partial \frac{1}{2} e^2(k)}{\partial \boldsymbol{S}} &= e(k) \frac{\partial e(k)}{\partial \boldsymbol{S}} \\ &= \Phi(k) \left(\boldsymbol{S}^T \Phi(k) - r(k) \right), \end{aligned} \tag{7.27}$$

where

$$\Phi(k) = \boldsymbol{\varphi} \left(\mathbf{X}_R(k), \boldsymbol{u}_R(k) \right) - \gamma \boldsymbol{\varphi} \left(\mathbf{X}_R(k+1), \boldsymbol{u}_R(k+1) \right). \tag{7.28}$$

The parameter set $\boldsymbol{S}$ is then recursively updated until convergence, following:

$$\boldsymbol{S}^+ \leftarrow \boldsymbol{S} - \boldsymbol{l} \otimes \left(\Phi(k) \left(\boldsymbol{S}^T \Phi(k) - r(k) \right) \right), \tag{7.29}$$

where the vector $\boldsymbol{l} > 0$ represents the step size. It's value needs to be tuned during implementation.

Note that the parameter vector $\boldsymbol{S}$ can also be determined using recursive least square (RLS) method. A necessary condition for the convergence of $\boldsymbol{S}$ is that the regressor $\Phi(k)$ is persistently excited (PE) [156], namely:

$$\text{rank} \left(\Phi(1), ..., \Phi(k) \right) = \text{dimension of } \boldsymbol{S} \tag{7.30}$$

In practice, the PE condition can be satisfied by modulating the control input $\boldsymbol{u}$, e.g., generating a particular trajectory for $\boldsymbol{u}$ [156] or adding an appropriate probing noise [140]. However, both methods are critical in the context of human-robot collaboration. Firstly, the control input is usually task-related and therefore cannot be modified arbitrarily. Secondly, adding a probing noise might reduce human comfortableness and cause problems in robot control (e.g., chattering).

As long as the minimum of Q-function is reached, the optimal control input $\boldsymbol{u}_R^\star$ is determined by solving:

$$\frac{\partial Q^* \left(\mathbf{X}_R, \boldsymbol{u}_R \right)}{\partial \boldsymbol{u}_R^*} \equiv \boldsymbol{0} \tag{7.31}$$

It yields:

$$\boldsymbol{u}_R^* = -\boldsymbol{S}_{uu}^{*-1} \boldsymbol{S}_{xu}^{*T} \mathbf{X}_R. \tag{7.32}$$

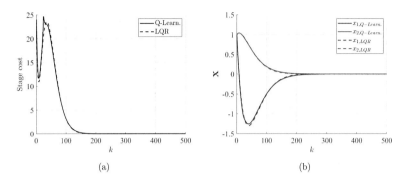

(a)   (b)

Figure 7.3: Comparison of two controllers based on Q-learning and LQR: (a) stage cost, (b) system states.

Numerical example

Consider following discrete-time linear system:

$$\begin{pmatrix} x_1(k+1) \\ x_2(k+1) \end{pmatrix} = \begin{pmatrix} 0.95 & -0.1 \\ 0.01 & 1 \end{pmatrix} \begin{pmatrix} x_1(k) \\ x_2(k) \end{pmatrix} + \begin{pmatrix} 0.01 \\ 0 \end{pmatrix} u(k), \quad x_1(0) = x_2(0) = 1. \quad (7.33)$$

The weighting matrices in the cost function are selected as $\boldsymbol{Q} = \begin{pmatrix} 10 & 0 \\ 0 & 10 \end{pmatrix}$ and $R = 1$.

The system is controlled by a LQR controller and an adaptive controller based on Q-learning. For Q-learning, a probe noise is added in the control input to guarantee PE, and the discount factor γ is chosen as 0.9. Fig. 7.3 shows that the performance of the Q-learning based controller is close to the LQR-controller. The cumulative costs are 1517.4 (for Q-learning) and 1498.5 (for LQR). The simulation indicates that the Q-learning based controller yields a near-optimal solution.

Relation to impedance control

Rewriting Eq. (7.32) as:

$$\boldsymbol{u}_R^* = \boldsymbol{K}_d^\star \left(\boldsymbol{p}_{R,d} - \boldsymbol{p}_R \right) - \boldsymbol{D}_d^\star \dot{\boldsymbol{p}}_R. \quad (7.34)$$

where $\boldsymbol{K}_d^\star, \boldsymbol{D}_d^\star$ are sub-matrices of $\boldsymbol{S}_{uu}^{\star-1} \boldsymbol{S}_{xu}^{\star T}$. The above PD type controller can be regarded as a mechanical spring-damper system, which generates an elastic force to cancel the position tracking error and achieve sufficient damping. This interpretation is the same with a Cartesian impedance controller. Hence, the Q-learning-based algorithm is equivalent to adaptive impedance control, and learning the optimal control policy means finding the optimal impedance parameter set.

Note that the desired mass M_d can be arbitrarily chosen since it would not change the learning process. For simplification, in this work, it is set to a unit matrix, namely $M_d = I$. Recalling the Cartesian impedance control law without force feedback in Eq. (6.12), the joint torque input is determined by:

$$\boldsymbol{\tau} = \boldsymbol{M}(\boldsymbol{q})\boldsymbol{J}^{-1}(\boldsymbol{q}) \left(\boldsymbol{K}_d^* \left(\boldsymbol{p}_{R,d} - \boldsymbol{p}_R\right) + \boldsymbol{D}_d^* \dot{\boldsymbol{p}}_R, -\dot{\boldsymbol{J}}(\dot{\boldsymbol{q}}, \boldsymbol{q})\dot{\boldsymbol{q}} \right)$$
$$+ \, \boldsymbol{C}(\dot{\boldsymbol{q}}, \boldsymbol{q})\dot{\boldsymbol{q}} + \boldsymbol{g}(\boldsymbol{q}), \tag{7.35}$$

where $\boldsymbol{q}$ is the vector of joint angles, $\boldsymbol{M}$ is the robot inertia matrix, $\boldsymbol{C}$ is the matrix describing centrifugal and Coriolis terms, $\boldsymbol{g}$ is the vector of gravitational moments, $\boldsymbol{J}$ is the Jacobian matrix that describes the velocity kinematic.

Stability and optimality

The proposed Q-learning-based adaptive impedance control has a hierarchical structure: a low-level Cartesian impedance controller regulating the robot end-effector's dynamic and a high-level learning algorithm to optimize the impedance control gains.

Firstly, the stability of the low-level impedance control has been well-defined in literature. As long as $\boldsymbol{K}_d^*$ and $\boldsymbol{D}_d^*$ are positive definite, the low-level impedance control loop is stable [52]. A sufficient condition is that the learned parameters $\boldsymbol{S}_{xx}, S_{xu}$ and $\boldsymbol{S}_{uu}$ must be positive definite as well.

Moving to the upper level, according to the precious study in [157], the convergence of Q-learning is guaranteed if: (1) the system (described by $\boldsymbol{A}_d, \boldsymbol{B}_d$) is stabilizable, (2) the initial control policy $\boldsymbol{\mu}_0$ stabilizes the system, (3) the vector $\boldsymbol{\Phi}$ is persistently excited. Note that these are not the only conditions for convergence of Q-learning based on linear function approximation. Different conclusions can be found in literature, e.g., [158]. Moreover, one related work [159] points out that Q-Learning as well as its convergence proof is only valid in discrete-time systems and cannot be directly transformed into continuous-time systems.

Authors in [157] have concluded that if all the conditions mentioned above are satisfied, the Q-learning converges to the optimal control policy. However, it is an ideal case that is almost unachievable in practice. With consideration of function approximation errors, the approximate Q-learning only gives a near-optimal solution [160, 155]. A quantitative analysis of the error bonds has been performed in [160] and can be used to design robust Q-learning approaches.

In summary, besides the convergence and optimality of function-approximation-based Q-learning, the stability of the adaptive impedance control as a whole is still an open issue and should be investigated in future works.

Handling the contact force

As presented at the beginning of this section, the impedance parameters are determined by solving the local optimization problem on the robot's side. Hence, there is no guarantee that they can completely satisfy the human's expectations. If not, the human will generate extra control effort to correct the object motion, which increases the contact force.

As discussed in a previous study [161], interaction force contributes significantly to the sensory communication in PHRC and can be used as a measurement of "disagreement". If the human collaborator keeps applying large force, it is probably because they intend to correct the current motion of the object. Based on this assumption, a further adjustment of the impedance parameters, named "negotiation", is performed. The aim is to decrease the impedance parameters as the contact force increases, making the robot behaves more compliantly. From both physical and psychological points of view, acceptable forces are crucial at first glance in PHRC [162]. Hence, it is critical if the robot generates a large resistant force on the human.

Under this concept, for each relevant DOF, a discount factor α_i is defined as follows:

$$\alpha_i = \begin{cases} 1 - \frac{\|f_i\|}{f_{t,i}}, & \text{if } \|f_i\| < f_{t,i} \\ 0. & \text{else} \end{cases} \tag{7.36}$$

where f_i represents the interaction force on i-th DOF, and the force threshold $f_{t,i} > 0$ is a design parameter. Then the robot stiffness is modified as:

$$\hat{K}_{d,i} = \left(K_{d,i}^*\right)^{\alpha_i}. \tag{7.37}$$

The new damping is computed by:

$$\hat{D}_{d,i} = D_i^* \sqrt{\frac{\hat{K}_{d,i}}{K_{d,i}^*}}, \tag{7.38}$$

so that the damping ratio of the equivalent second-order system remains unchanged to avoid oscillations caused by under-damping. Note that if the threshold $f_{t,i}$ is exceeded, α_i becomes zero, and the stiffness turns to 1, which results in very compliant behavior.

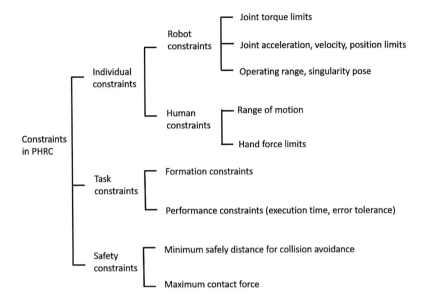

Figure 7.4: An overview of constraints in PHRC.

7.3.2 Learning with consideration of constraints

So far, the Q-learning-based adaptive impedance control has been discussed without consideration of constraints. As reviewed in [153], constraints handling remains immature in RL, except for input constraints. Direct handling of state constraints is still an open issue. In most constrained-RL algorithms, state constraints are only treated as soft constraints, namely by adding a penalty function for constraints violation to the objective function.

Fig. 7.4 briefly summarizes the constraints in the context of PHRC. It is impractical to consider all of them in the learning process. Some constraints such as operating range, formation and performance constraints, can be included in the trajectory planning. Robot joint constraints, e.g., toque limits, maximum velocity, are usually integrated into the low-level joint controller developed by the manufacturer. This section chooses two types of constraints that are essential for the impedance control , and discusses some possible approaches to considering constraints in the Q-learning algorithm.

The first issue relates to the ability to maintain the desired stiffness in Cartesian space under impedance control, which is named as "stiffness feasibility" in [163]. As introduced before, impedance control makes the robot behave like a mass-spring-damper system and generate an "elastic" force when interacting with the environment. According to the law of action and reaction, a reactive force with the same amplitude acts on the robot-effector and is mapped to joint space through the Jacobian matrix. However, the resulted torque may exceed the joint torque limits so that the robot cannot generate enough torque to maintain its position and stiffness simultaneously. To overcome this problem, it is advantageous to consider the stiffness feasibility in the learning phase so that the algorithm does not generate a stiffness reference value that the robot cannot achieve.

Firstly, the joint torques are divided into two parts: one for maintaining the desired impedance behavior in Cartesian space, denoted as $\boldsymbol{\tau}_{imp}$, the other for compensating the non-linear dynamics, denoted as $\boldsymbol{\tau}_{ff}$. It yields:

$$\boldsymbol{\tau} = \boldsymbol{\tau}_{imp} + \boldsymbol{\tau}_{ff}, \tag{7.39}$$

where

$$\boldsymbol{\tau}_{ff} = \boldsymbol{C}(\dot{\boldsymbol{q}}, \boldsymbol{q})\dot{\boldsymbol{q}} + \boldsymbol{g}(\boldsymbol{q}). \tag{7.40}$$

Next, define a scaling factor including N joint limits:

$$\boldsymbol{W} = \text{diag}\left(\frac{1}{\tau_{lim,1}} \quad \frac{1}{\tau_{lim,2}} \quad ... \quad \frac{1}{\tau_{lim,N}}\right). \tag{7.41}$$

Then the joint torques can be scaled as follows:

$$\boldsymbol{\tau}^{'} = \boldsymbol{W}\left(\boldsymbol{\tau}_{imp} + \boldsymbol{\tau}_{ff}\right). \tag{7.42}$$

The constraint for joint torque limits can be formulated with L$_2$ norm of $\boldsymbol{\tau}^{'}$, namely:

$$\left\|\boldsymbol{\tau}^{'}\right\| \leq 1 \Leftrightarrow \left(\boldsymbol{\tau}^{'}\right)^{T}\boldsymbol{\tau}^{'} \leq 1. \tag{7.43}$$

Transforming the above inequality into Cartesian space, it yields:

$$\left(\boldsymbol{f}_{imp}^{T}\boldsymbol{J}\left(\boldsymbol{q}\right) + \boldsymbol{\tau}_{ff}^{T}\right)\boldsymbol{W}^{T}\boldsymbol{W}\left(\boldsymbol{J}^{T}\left(\boldsymbol{q}\right)\boldsymbol{f}_{imp} + \boldsymbol{\tau}_{ff}\right) \leq 1 \tag{7.44}$$

As discussed before, the impedance force $\boldsymbol{f}_{imp}$ is equal to the control input $\boldsymbol{u}_{R}$. Denoting $\boldsymbol{H} = \boldsymbol{J}\left(\boldsymbol{q}\right)\boldsymbol{W}^{T}\boldsymbol{W}\boldsymbol{J}^{T}\left(\boldsymbol{q}\right)$, $\boldsymbol{k} = \boldsymbol{J}\left(\boldsymbol{q}\right)\boldsymbol{W}^{T}\boldsymbol{W}\boldsymbol{\tau}_{ff}$ and $d = \boldsymbol{\tau}_{ff}^{T}\boldsymbol{W}^{T}\boldsymbol{W}\boldsymbol{\tau}_{ff} - 1$, the constraints is reformulated with the control input:

$$\boldsymbol{u}_{R}^{T}\boldsymbol{H}\boldsymbol{u}_{R} + 2\boldsymbol{k}^{T}\boldsymbol{u}_{R} + d \leq 0 \tag{7.45}$$

The above equation can be integrated into the TD-Learning problem, it yields:

$$\min_{\boldsymbol{s}} \frac{1}{2}e^{2}(k)$$

$$\text{s.t.} \quad e(k) = -\boldsymbol{S}^T \boldsymbol{\varphi} \left(\mathbf{X}_R(k), \boldsymbol{u}_R(k)\right) + r\left(\mathbf{X}_R(k), \boldsymbol{u}_R(k)\right)$$
$$+ \gamma \boldsymbol{S}^T \boldsymbol{\varphi} \left(\mathbf{X}_R(k+1), \boldsymbol{u}_R(k+1)\right),$$
$$\boldsymbol{S} = \left(\text{vec}\left(\boldsymbol{S}_{xx}\right) \quad \text{vec}\left(\boldsymbol{S}_{xu}\right) \quad \text{vec}\left(\boldsymbol{S}_{uu}\right)\right)^T,$$
$$\boldsymbol{S}_{xx}, \boldsymbol{S}_{xu}, S_{uu} \succ 0,$$
$$\boldsymbol{u}_R^T \boldsymbol{H} \boldsymbol{u}_R + 2\boldsymbol{k}^T \boldsymbol{u}_R + d \leq 0,$$
$$\boldsymbol{u}_R = -\boldsymbol{S}_{uu}^{-1} \boldsymbol{S}_{xu}^T \mathbf{X}_R(k). \tag{7.46}$$

Solving this constrained optimization problem online results in a control policy that minimizes the TD error and is achievable for the low-level impedance controller. Note that the inverse term $\boldsymbol{S}_{uu}^{-1}$ may cause some difficulties when solving the problem. One possible solution is to set $\boldsymbol{S}_{uu}^{-1}$ as a constant since what matters is the product $\boldsymbol{S}_{uu}^{-1} \boldsymbol{S}_{xu}^T$. However, the function approximation will thereby lose some DOFs. How large the influence can be on the approximation error remains unclear.

The stiffness feasibility still belongs to input constraints. Directly handling state constraints in model-free Q-learning is difficult. On the other hand, it is critical since the learned control policy can still generate undesired robot behaviors even with consideration of input constraints. One intuitive solution would be to build a second model-based optimization problem after Q-learning, namely:

$$\min_{\hat{\boldsymbol{u}}_R(k)} \frac{1}{2} \left\| \hat{\boldsymbol{u}}_R(k) - \boldsymbol{u}_R^*(k) \right\|^2$$
$$\text{s.t.} \quad \mathbf{X}_R(k+1) = \boldsymbol{F}\left(\mathbf{X}_R(k)\right) + \boldsymbol{B}\left(\mathbf{X}_R(k)\right) \hat{\boldsymbol{u}}_R(k) \quad \text{(robot dynamic)},$$
$$\mathbf{X}_R(k+1) \in \boldsymbol{\mathcal{X}}, \quad \text{(state constraint)}. \tag{7.47}$$

The above optimization problem refers to constrained-least-square, which minimizes the distance between the learned "optimal" control input through Q-learning and the actual control input sent to the robot. Solving the optimization problem is not complex since it only requires a one-step prediction of the robot's state. The result is a near-optimal solution.

Numerical example

To validate the two-phase Q-learning-LS method introduced above, the system in Eq. (7.33) is again used for simulation. The weighting matrices and the parameters of the Q-learning algorithm are also the same. Twp state constraints $-1 \leq x_1(k) \leq 1$, $-2 \leq x_2(k) \leq 2$ are added. For comparison, a controller optimized based on constrained quadratic programming is investigated. The results in Fig. 7.5 show that the constraints are held via both controllers. The cumulative costs are 1866.0 (for Q-Learning-LS) and

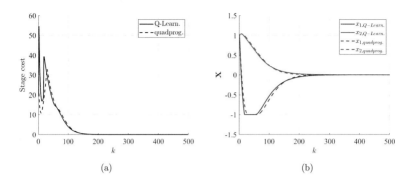

Figure 7.5: Comparison of two controllers based on Q-learning-LS and quadratic programming for a constrained optimization problem: (a) stage cost, (b) system states.

1675.3 (for quadratic programming). The simulation indicates that the Q-learning-LS method can achieve near-optimal performance.

7.3.3 Impedance and role assignment

As generally known, humans and robots have complementary capabilities. For instance, robots have a flair for precision and repeatability, while humans are good at analysis and decision making. In order to utilize the capabilities of both sides, a proper role assignment and adaption strategy is of great importance and has been investigated by researchers from various fields, e.g., psychology, cognitive science, and robotics [164].

Leader-follower (or master-slave) is still one commonly implemented strategy for PHRC in which the human dominates the collaboration as the leader, and the robot should follow all the commands or motion trajectories from the human. However, this scheme is far from optimal since: 1) extra human effort is required for task coordination. 2) the robot only works "passively" under human guidance and cannot make any proactive contribution. 3) sometimes, the robot can even make better decisions, especially when the sense of human is interrupted. Hence, recent researches [165, 166, 167] suggest a dynamic role allocation/adaption strategy, i.e., the roles of leader/follower can be adjusted during the collaboration.

In PHRC, the role adaption is signalized through the contact force. For example, the human starts to dominate the collaboration by applying a large force. Then both the human and the robot negotiate by varying their impedance parameters. If the one agent accepts the change of role assignment, it will reduce its impedance value and behave compliantly as a follower, and vice versa. Hence, impedance is a crucial aspect of role adaption in physical interactions, for both human-human and human-robot. Learning the optimal impedance parameters helps to achieve an optimal role distribution as well.

7.4 Summary

This chapter presented how reinforcement learning can solve an infinite-horizon optimization problem and how to combine the solution with robot impedance control. Based on Bellman's principle of optimality, the Q-learning method studied in this work is built upon dynamic programming. Q-learning provides a model-free algorithm that learns the optimal value function online through function-approximation techniques instead of analytically solving the optimization problem based on the system dynamic model. The control policy is then adapted along the negative gradient of the Q-function.

The critical factors for the combination of Q-learning with adaptive impedance control are: (1) modeling of the interaction dynamic through coupled mechanical impedances, (2) parameterizing of the reward function in a quadratic form. Under these conditions, the optimal value function should also be quadratic, and the optimal polity, determined by the derivative of the value function, is linear. Thereby, the control input can be physically interpreted as an elastic force generated by a virtual spring-damper system to cancel the motion error. The stiffness and damping are equal to the gain matrices.

The contact force plays an essential role in PHRC. In this work, it is regarded as a measurement of human disagreement, i.e., when the robot senses increasing contact force on its end-effector, it means that the human intends to correct the robot motion. Based on this idea, the impedance parameters determined by RL are further adjusted based on the contact force feedback.

Constrained RL has been drawn more and more attention in recent years for real-world robotic applications. Unconstrained learning might result in policies that are unachievable or lead to dangerous states. This work discusses possible extensions of the proposed control method with consideration of constraints. A comprehensive study should be included in future work.

In the next chapter, the proposed Q-learning-based adaptive impedance control will be combined with human motion prediction and tested in experiments.

8 An Adaptive Learning and Control Framework in Human-Robot Collaboration

8.1 Introduction

This chapter brings the content of all the previous chapters together, presents an adaptive learning and control framework, and validates it experimentally through different human-robot collaboration tasks. The framework is built on the basis of Cartesian impedance control. The "learning" part includes offline learning of human motion patterns as well as online learning of an optimal control policy to track the reference trajectory and interact with human. The "adaption" part contains the adaption of the human model, the reference trajectory, and the impedance control parameters for the robot. In the experimental study, two typical benchmark applications, object-handover and object-handling are chosen to validate the proposed framework in both contact-free and physical interactions. Various learning and adaptation methods are tested and evaluated.

Fig. 8.1 gives a general overview of content in this chapter. Each color represents one approach. The blue one is a GP-based method in which the human motion is learned and predicted by online GP. The prediction output is used for robot reference trajectory planning. Moreover, the impedance control parameters are adjusted based on the uncertainties determined by GP regression. The green one is a DMP-GP-based method. A DMP model is learned for the description of human motion, in which the nonlinear shape-attraction term is represented by GP. During the online execution phase, the model parameters are adapted according to human motion prediction error feedback. Both methods are validated in a human-robot handover task. More details can be found in Section 8.2. The orange color represents the Q-learning-based adaptive impedance control approach. Moreover, to make the robot flexible to the change of human's intention, a human reference estimator based on a simple linear contact stiffness model is proposed. The experimental study of this approach is presented in Section 8.3.

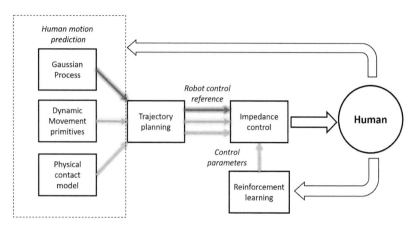

Figure 8.1: An overview of the approaches discussed in this chapter.

This chapter also presents some unique designs which aim to solve several practical problems, e.g., variation handover locations, unknown human targets, etc. The performance of all experiments is evaluated by both subjective (comfortableness, fluency) and objective metrics (tracking error, execution time, force).

8.2 Experimental study: handovers

This thesis considers the scenario that a human passes an object to a robot and focuses on motion planning and control in the pre-handover phase, in which the giver and the receiver move towards each other to get close enough to transfer the object. The main task is to generate safe and human-aware robot trajectories. The robot should be able to adjust its motion online according to human feedback with a similar movement profile to human hand. Since no physical contact takes place in the reaching phase, the impedance control works as a pure motion controller to track the reference trajectories.

Two designs have been made to achieve an on-the-fly handover, i.e., the robot should move simultaneously with the human. The first design follows a "prediction + tracking" strategy. Firstly, Gaussian process (GP) is used to predict human hand motion for a particular look-ahead time. Afterward, the robot generates a point-to-point motion trajectory by taking the predicted human hand position as the target. Both motion prediction and trajectory planning are updated online according to human motion feedback. The second design follows a "learning + adaption" strategy. The robot first learns how

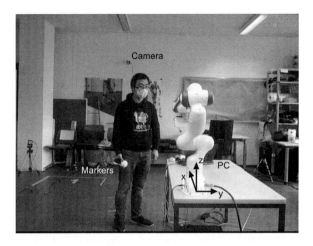

Figure 8.2: An overview of the experimental platform

the human performs a handover task through dynamic movement primitives. Then in the online execution phase, the learned DMP model is used to predict human motion and generate robot trajectories. The model parameters are adapted based on the prediction error.

8.2.1 Hardware

All experiments presented in this thesis were preformed with a 7-DoF *Franka Emika* robot arm equipped with a 2-finger gripper. The robot motion in Cartesian coordinates was determined based on the measured joint data with a sampling rate of 1 kHz and its kinematic model. The control algorithm was implemented in C++ under Ubuntu 16.04 using the *Franka Control Interface* and executed on an external PC (Intel Core i7-3530M (2.9-3.6 GHz), 8 GB RAM) connected via Ethernet. The human hand motion was tracked by an *OptiTrack* motion capture system with a sampling frequency up to 200 Hz. The markers were fixed on the back of the human hand. Another PC with software *Motive* was used for real-time measurement data processing and streaming. The communication between both PCs took place through Ethernet as well. Since *OptiTrack* only delivers a measurement of positions, the corresponding velocities were estimated by a Kalman-filter with the constant velocity model. An overview of the experimental platform is shown in Fig 8.2.

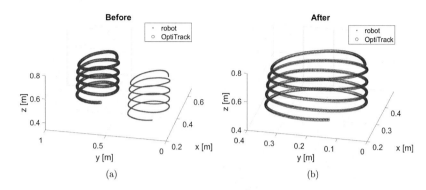

(a)   (b)

Figure 8.3: Pre-alignment of the two coordinate frames with measured position data. (a) original point sets of the same trajectory before alignment (red: in robot coordinate frame, blue: in *OptiTrack*). (b) point sets after alignment, both in robot coordinate frame.

8.2.2 Pre-alignment

Measurements acquired by an optical motion capture system are usually carried out in a fixed coordinate frame. One issue in this work is that the training data for offline learning a GP model was recorded in a different coordinate system from the one used for the online prediction. Moreover, the *OptiTrack* system also has a different coordinate frame as the one used in the robot. Hence, it is necessary to define a joint reference frame and transform all different coordinate frames into it. For this purpose, the *rigid Extended Coherent Point Drift* (rECPD) method proposed in [168] was used.

Firstly, the coordinate systems of the robot and *OptiTrack* were aligned. The procedure was as follows: firstly, a rigid body (a marker set consists of 4 markers) was fixed at the fingertips of the robot gripper. Then the robot executed a pre-defined 3D-trajectory, and its end-effector position was recorded by both *OptiTrack* and its sensor system. Afterwards, the two measurement sets were aligned using rECPD, and the transformation matrix between the two coordinate frames was calculated. The data sets before and after alignment of the same trajectory were shown in Fig 8.3. The root-mean-square error of the alignment was 2.1 mm.

Secondly, the coordinate systems for training and validation were aligned. The procedure was the same: at first, the human participant was asked to perform the same motion as in the training phase several times, then the data was taken as the input of the rECPD algorithm for computation the required parameters for the alignment. Note that, unlike a robot, a human cannot reproduce an equivalent motion as before, even for the same type of motion. The error of the alignment is thereby larger. In this work, it was taken as a remedy since the calibration information of the motion capture system for recording the training data was missing.

8.2.3 Design based on Gaussian Process

Fig. 8.4 shows an overview of the system module. The design is made based on the online-sparse GP regression method presented in Chapter 3, Section 3.3. The offline training phase includes learning of the sparse GP model and the pre-alignment of different coordinate systems. The online execution phase consists of a trajectory generation thread and a motion control thread, with a sampling period of 50 ms and 1 ms respectively. The reference trajectory is tracked by a Cartesian impedance controller. The controller parameters are online adjusted by a "danger index" which relies on prediction uncertainties and the relative distance between the human hand and the robot.

Trajectory planning

As presented in Chapter 3, the GP-prediction produces a sequence of human hand positions based on the current measurement, namely:

$$\boldsymbol{x}_H(k) \rightarrow (\hat{\boldsymbol{x}}_H(k+1), ..., \hat{\boldsymbol{x}}_H(k+N))^T$$

Then the last point $\hat{\boldsymbol{x}}_H(k+N)$ is taken as a virtual target for the robot. The value should be transformed into the robot coordinate system by multiplying the homogeneous transformation matrix determined from the pre-alignment:

$$\boldsymbol{x}_T(k) = \boldsymbol{R}_H^R \cdot \hat{\boldsymbol{x}}_H(k+N) + \boldsymbol{l}_H^R, \tag{8.1}$$

where $\boldsymbol{R}_H^R$ represents the rotation matrix of the *OptiTrack* coordinate frame with respect to the robot frame, $\boldsymbol{l}_H^R$ is the vector between the origin of the two coordinate frames.

After that, a point-to-point motion between the current robot position $\boldsymbol{x}_R(k)$ and the virtual target $\boldsymbol{x}_T(k)$ is generated as a polynomial function of time. Motivated by the minimum jerk model introduced in Chapter 2, in each dimension, the reference trajectory is parameterized by a fifth-order (quintic) polynomial contains six coefficients determined by six constraints, i.e., the initial and final positions, velocities, and accelerations. In each iteration, the current robot motion data is taken as the initial condition. The

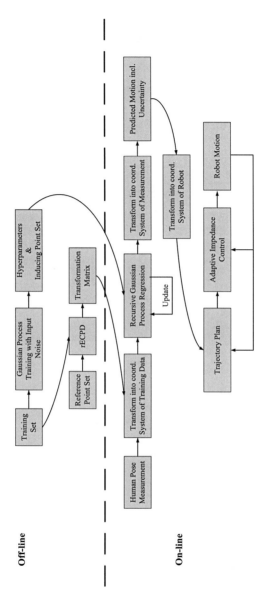

Figure 8.4: Overview of the proposed system for human motion prediction and impedance control based on online GP regression.

virtual target is defined as the final position. Terminal velocity and acceleration are both set to zero.

Another parameter that needs to be specified is the total travel time of the trajectory, which decides how fast the robot moves towards its target. According to the feedback of labor participants, it is concluded that it is more acceptable for the human if the robot moves almost at the same speed as human. Hence, the total travel time, denoted as T_f, is adjusted by taking human's velocity into account, it yields:

$$T_f(k) = \frac{\|\boldsymbol{x}_T(k) - \boldsymbol{x}_R(k)\|}{\|\bar{\boldsymbol{v}}_h\|}, \tag{8.2}$$

where $\|\bar{\boldsymbol{v}}_h\|$ is average of human velocity observed from human demonstrations. In each iteration, N reference values are generated for robot control. The parameters for trajectory planning are updated when new predictions are available.

Robot Control

Safety is the most significant factor in human-robot interaction scenarios and can be handled by planning and control strategies. One example of safe planning techniques can be found in [36], where a human was regarded as a dynamic obstacle, and the robot should adjust its motion to avoid collisions. However, this approach is not suitable here since, for a handover task, the robot should track the human hand instead of avoiding it. In this experiment, the safety criterion is considered in the control part. Firstly a danger index d_k of interaction is defined as follows:

$$d(k) = a_1 \frac{1}{\|\boldsymbol{x}_H(k) - \boldsymbol{x}_R(k)\|} + a_2 \max \boldsymbol{\Sigma}_*(k), \tag{8.3}$$

where $a_1, a_2 > 0$ are design parameters which represent weightings of the two terms, $\boldsymbol{\Sigma}_*$ is the variance matrix of the GP prediction. The first term is the inverse of the distance between the human and the robot. If the robot gets close to the human, the risk of collision arises. The second term relies on the maximum value of the covariance matrix of the predicted human motion, which can be regarded as a measurement of uncertainty. If the value is high, it means the robot is quite uncertain about its prediction of human motion. It is usually because humans perform some unnatural motion that has not been seen by the robot before. This would also be regarded as a dangerous situation.

The danger index d_k is then used to adapt the stiffness matrix in the impedance control, namely:

$$\hat{\boldsymbol{K}}_d = \begin{cases} \boldsymbol{K}_d(1 - \gamma d_k^2), & \text{if } \gamma d_k^2 < 1 \\ \boldsymbol{0}. & \text{else} \end{cases} \tag{8.4}$$

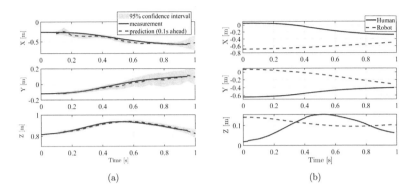

Figure 8.5: Experimental results: (a) comparison of the prediction (dash) and real measurements (solid) in respect to coord. system of the training set. (b) motion trajectory of human (solid) and robot (dash) overtime.

$\gamma > 0$ is a tuning parameter, which describes how fast the stiffness decreases according to the danger index. From the physical point of view, the desired end-effector behavior of the robot is effectively a spring-damper system. Hence, reducing the stiffness of the spring will make the system dynamic slower. As a result, the movement of the robot will slow down. In the extreme case, the stiffness becomes zero, and the robot is decelerated by the remaining damping factor D_d and other damping effects until standstill.

Results

Fig. 8.5(a) shows the performance of the prediction of human motion. The look-ahead time was set to 100 ms, i.e., approximately 10% of the whole trajectory. The predicted values are reliable with RMSE less than 1 cm. Additionally, compared to the training data, the human behaved slightly differently when the robot was present during the experiment. Hence, it would be interesting to study how a robot can influence the behavior of a human in an interaction scenario.

Adaption of the controller parameter based on the evolution of the danger index is plotted in Fig. 8.6. In the beginning, the danger index is dominated by the uncertainty of the prediction. Then its value increased rapidly since the robot was getting close to the human, and their distance became the deciding factor. A high value of danger index leads to a decrease of the stiffness in impedance control, which makes the robot's end-effector more compliant.

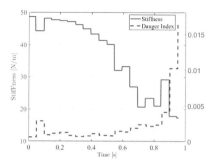

Figure 8.6: Adaption of controller parameter (solid) based on the danger index (dash).

Fig. 8.5(b) presents the trajectories of both the human and the robot over time. It can be seen that the robot started immediately as the human moved, and the motion trajectories were smooth. On the other hand, there still existed some distance between the human and the robot when they were supposed to meet each other. It led to extra human effort to move further and put the object on the robot's gripper. The reason is that the stiffness of the impedance controller is not large enough to eliminate control errors. This can be improved by fine-tuning the adaptive law to keep some stiffness, especially in the end phase.

Discussion

This experiment was the first framework implementation that included human motion prediction, online trajectory planning, and robot control. Results show that the basic functionalities are achieved: (1) The GP-based predictor delivers reliable online human motion prediction. (2) The robot's reaching motion is smooth and without delay. (3) The adaption of the impedance parameter based on prediction uncertainties is validated. Overall, the handover process is fluent and safe.

However, it should be pointed out this study has many limitations. Some of them rely on the algorithm. First, if the handover configurations (e.g., initial/goal positions, human speed) vary, the predictor requires new data to retrain the model, making the framework less efficient and flexible. As discussed in Chapter 3, it is mainly due to the zero-mean assumption of the GP model. Secondly, the adaption of the impedance parameters results in a conservative motion, and sometimes the robot stops too early. On the other hand, the experiment lacked solid performance evaluation. The number of repetitions is not large enough for statistical analyses. A comprehensive experimental study on the advantages of the proposed method in comparison to other state-of-the-art approaches should have been included. All these issues have been considered in the following experimental study.

8.2.4 Design based on Dynamic Movement Primitives

This section presents the experimental study of a DMP-inspired learning and control framework with application to human-robot handovers. The framework combines DMP with GP, i.e., using GP to fit the shape attraction term in DMP. The theoretic part was introduced in Chapter 4. As an extension of the previous design, the DMP-based method should be able to deal with variation handover locations and different human movement types.

An overview of the system is shown in Fig. 8.7. The framework includes two coupled DMPs, one for human motion prediction as giver and the other for robot trajectory generation as receiver. Both DMP models have the same mathematical formulation described in Eq. (4.14) but with different spatial scaling factors. The DMP models are first learned offline by several human demonstrations and then adapted online based on the human prediction error. A Cartesian impedance controller is used for tracking the reference trajectory generated by the robot DMP model. This design focuses on the adaption of the reference trajectory. The impedance parameters are kept constant.

Online adaption of the DMP model

As discussed in [48], the parameters in DMP can be categorized into three types: (1) parameters of the linear term (K and D), (2) parameters in the non-linear term (here the kernel function of GP), and (3) the time scaling τ and goal position g. Although all these parameters can influence the system dynamic described by DMP, authors in [48] pointed out that for a particular behaviour, the parameters in (3) play an essential role in trajectory modulation. In handover tasks, since the execution time normally varies not much, the goal position, namely the handover location becomes the most crucial parameter for online trajectory generation.

One practical issue is that the handover location is typically unknown before motion starts. On the other hand, it should not be fixed a priori. Otherwise, the human must adapt to the robot [12]. Ideally, the robot should consistently estimate the intended handover location of human and thereby adjust its motion online. Several previous works aimed to predict the handover location using model-based state estimation ap-

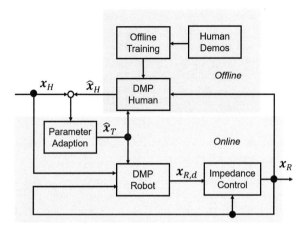

Figure 8.7: The structure overview of the proposed DMP inspired controller. x_H: human motion measurement, $\hat{x}_H$: human motion prediction, $\hat{x}_T$: estimated handover location, $x_{R,d}$: robot reference trajectory, x_R: robot end-effector motion.

proaches [11, 169, 111]. Their results show that a precise prediction of handover location is challenging, especially in the early phase of the movement. In contrast, authors in [110] argued that it was not necessary to estimate the handover location within the concept of DMP, if one directly takes the human hand position as a moving goal. However, their nonlinear term was not scaled by the goal position. As shown in Chapter 4, Section 4.5, this concept also has problems.

In this work, both ideas mentioned above are combined together in one framework. The actual human position and velocity are set as reference values in the linear feedback term. At the same time, the predicted handover location is adjusted online to correct the spatial scaling of the nonlinear shape-attraction term. This concept has two advantages: (1) the prediction error in the handover location does not influence the tracking behaviour in the linear term, (2) the non-linear feedforward term can be online adjusted if the handover location varies, since it will influence the spatial scaling. This is especially important in the early handover phase, in which the non-linear term dominates the motion. Note that the actual human hand position should not be directly used to correct the spatial scaling factor because it makes the nonlinear term too sensitive to human motion variation.

A simple method to adapt the estimated handover location is designed based on human motion prediction error. Substituting g, v_g in Eq. (4.14) with robot position x_R and velocity v_R, a DMP model for human motion prediction is obtained. Note that for calculating the scaling factor γ and $\boldsymbol{R}$ (Eq. (4.8)-(4.10)), the estimated handover location, denoted as $\hat{\boldsymbol{x}}_T$, is used instead of $\boldsymbol{x}_R$ as the target position $\boldsymbol{g}$. This DMP model generates a predicted human position $\hat{\boldsymbol{x}}_H$, and the value is compared with the measurement $\boldsymbol{x}_H$. The prediction error is denoted by $\boldsymbol{e} = \boldsymbol{x}_H - \hat{\boldsymbol{x}}_H$. Assuming that this error is caused by inaccurate estimation of the handover location $\hat{\boldsymbol{x}}_T$, a gradient-type update law is defined as follows:

$$\dot{\hat{\boldsymbol{x}}}_T = -\boldsymbol{P}\boldsymbol{e}, \qquad (8.5)$$

so that the change rate of $\hat{\boldsymbol{x}}_T$ is proportional to the negative gradient of the squared error $\boldsymbol{e}^T\boldsymbol{e}$. $\boldsymbol{P} \succ 0$ represents a diagonal gain matrix.

Then another DMP model is built to generate reference trajectory for robot, in which the human motion $\boldsymbol{x}_H, \boldsymbol{v}_H$ is considered as target position and velocity. Similarly, together with the initial robot position, the estimated handover location $\hat{\boldsymbol{x}}_T$ is used for computing the scaling parameters γ and $\boldsymbol{R}$ for the non-linear term.

Algorithm

The whole procedure, including human motion prediction, handover location estimation, robot trajectory planning and robot joint torque references generation, is summarized in Algorithm 1.

Algorithm 1 DMP-GP based controller

Require: Human measurement $\boldsymbol{x}_H, \boldsymbol{v}_H$, robot measurement $\boldsymbol{x}_R, \boldsymbol{v}_R$, hyperparameters of GP, parameters of DMP, initial guess of handover location $\hat{\boldsymbol{x}}_{T,0}$

1: **while** $\|\boldsymbol{x}_R - \boldsymbol{x}_H\| \geq$ offset **do**
2:   Calculate phase variable s according to Eq. (4.2)
3:   Calculate $\boldsymbol{f}(s)$ using GP regression according to Eq. (3.21), (3.22)
4:   Calculate $\omega(\tilde{\boldsymbol{x}})$ according to Eq. (4.11), (4.29)-(4.31)
   ▷ *Human motion prediction*
5:   Calculate scaling vector $\gamma, \boldsymbol{R}$ using $\hat{\boldsymbol{x}}_T, \boldsymbol{x}_{H,0}$ according to Eq. (4.8)-(4.10)
6:   Predict human position $\hat{\boldsymbol{x}}_H$ according to Eq. (4.14), using $\boldsymbol{x}_R, \boldsymbol{v}_R$ as $\boldsymbol{G}, \boldsymbol{v}_g$
7:   Update $\hat{\boldsymbol{x}}_T$ according to Eq. (8.5)
   ▷ *Robot control*
8:   Calculate scaling vector $\gamma, \boldsymbol{R}$ using $\hat{\boldsymbol{x}}_T, \boldsymbol{x}_{R,0}$ according to Eq. (4.8)-(4.10)
9:   Calculate robot reference $\boldsymbol{x}_{R,d}$ according to Eq. (4.14), using $\boldsymbol{x}_H, \boldsymbol{v}_H$ as $\boldsymbol{g}, \boldsymbol{v}_g$
10:   Calculate joint torque reference according to Eq. (6.30)
11: **end while**

Experiment

In all experiments, the human operator acted as a giver to pass an object to the robot. Each trial consisted of the following procedures:

1. Initialization: the participant held the object and put his/her hand naturally on the same side of the body. Both human and robot remained standstill. Their hand positions were recorded as the initial position x_0, which was used to initialize the scaling factors γ, R (Eq. (4.8)-(4.10)) and the estimated handover location (Eq. (8.5)).

2. Approaching: the participant brought the object to the robot. Simultaneously the robot started reaching the object. Its motion was triggered and adjusted by observing the position change of the human hand. No other communication signals such as gaze, verbal command, etc., are required.

3. Transfer: the participant released the object, then the robot grasped. Since no marker was installed on the object, and the robot gripper was not equipped with a force/tactile sensor, accurate grasping was challenging to achieve. Hence, one simplification was made. As long as the relative distance between the markers on the human hand and the robot end-effector was smaller than 5 cm (determined by the size of the object), the approaching phase was considered as finished then the robot started to grasp. Since the markers were fixed on the back side of human hand, not directly on the object, sometimes the robot can not locate the object accurately. In this case, the participant should offer help to put the object between the two fingers of the robot gripper.

4. Retreat: As grasping was finished, both human and the robot moved back to their initial positions.

Moreover, three different handover modes were defined:

1. Normal mode: The participant performed the handover in a natural and relaxed manner without paying much attention to speed, direction and meeting point. The aim was to test the basic functionality and the fluency of the proposed method.

2. Variation mode: The participant was asked to change the direction of movement during the approaching phase, caused by suddenly appeared obstacles or change of target position. The purpose was to evaluate the adaptability of the proposed method.

3. Passive mode: The participant performed a passive handover motion by reducing speed and travelling distance. This mode imitated the motion of injured patients

Handover Mode	without Correction	with Correction
Normal	100% (15/15)	93.33% (14/15)
Variation	60% (9/15)	93.33% (14/15)
Passive	73.33% (11/15)	100% (15/15)

Table 8.1: Success rate in different handover modes with- and without online correction

or elders. The goal was to test if the robot can generate proactive and assistive behaviour in such a case.

In this work, five participants took part in the experiment. In all experiments, each participant first chose a starting position arbitrarily, then performed all three handover modes to test one controller and repeated the process for the other one. After one round (6 trials), the participant should change its stand position. Totally 90 trials were recorded for quantitative analysis.

Moreover, all participants were interviewed after performing all the trials to provide feedback on user experience. In the interview, participants entered an evaluation score between 1 (unsatisfied) and 5 (very good) of two commonly used factors in human-robot interaction communities, namely comfortableness and fluency.

For comparison, a DMP-based controller without online correction was applied. The mathematical formulation is the same (see Eq. (4.14)). The only difference is that the factor γ and R were not updated online, but calculated based on a pre-specified constant transfer position, i.e., midway between human and robot.

Results

As reviewed in [12], a human-robot handover task can be evaluated by three types of metrics: (1) task performance metrics, (2) psychophysiological metrics, and 3) subjective metrics. In this work, since no psychophysiological measurement (e.g. electromyography, heart rate) was provided, only task performance and subjective metrics were considered.

One of the most commonly used performance indexes is the success rate, which performs a statistical view of the handover reliability. In this work, a successful handover should satisfy: (1) the robot end-effector stopped within a distance of 5 cm from the human hand. (2) the object exchange finished with successful grasping. The results in different handover modes with- and without online motion correction are listed in Tab. 8.1. It shows that the controller with online motion correction in the variation and passive modes lead to an improvement in the success rate. In both cases, the human motions are dynamic and abnormal. However, in the normal handover mode, the online correction performs slightly worse than the other one. In all failed attempts, the robot end-effector did not stop within the pre-specified distance from human hand. It is mainly caused by the prediction error in the handover location, which drives the robot in the wrong direction.

Fluency is another essential characteristic of human-robot interaction. As discussed in [170], it can be evaluated by both objective and subjective metrics. Objective metrics are usually determined by time measurements. In this work, the time of the reaching phase was measured, i.e. the time between motion start and the distance threshold (5 cm). The results are shown in Fig. 8.8(a). Similarly, in the variation and the passive modes, the controller with online motion correction reduces the reaching phase's travel-time by 13.77% and 17.39% respectively , thus increasing the fluency. In the normal mode, the average time is slightly longer (7.48%).

It should be noted that other time-relied metrics, especially human idle time and functional delay, are also highly correlated to fluency [170]. However, we did not consider them since their values were almost negligible in the reaching phase. In the experiments, the system idle mainly happened during the object exchange, in which both human and robot held their positions and waited until the grasping was finished. The idle time can be further reduced by designing a proper grasping configuration.

Moreover, fluency is also included as a factor in our user study as a subjective metric. As shown in Fig. 8.9, column 2, the controller with online correction received a 14.29% higher average score than the one without online correction .

Another widely used performance index in human-robot interaction is human effort, typically quantified by force/energy measurements. In this work, human effort is evaluated based on two factors: total kinetic energy and velocity trajectory smoothness. The former provides an overview of the energy consumption during the handover. The latter shows if the human needed to change its velocity frequently to adapt to the robot motion. For each successful trial, the quadratic sum of human hand velocity was calculated and regarded as a measurement of the total energy consumption. Smoothness was determined by calculating the arc length of the velocity profiles. According to the study in [171], the velocity profile with smoother movement should have a smaller arc length. The average values of the two factors in different handover modes are shown in Fig. 8.8(b) and Fig. 8.8(c). The result is similar to other performance indexes: in the variation and the passive mode, the controller with online correction reduces the velocity quadratic sum by 11.86% and 37.45% as well as the arc length of the velocity profile by 8.18% and 20.39% in comparison to the controller without online correction. However,

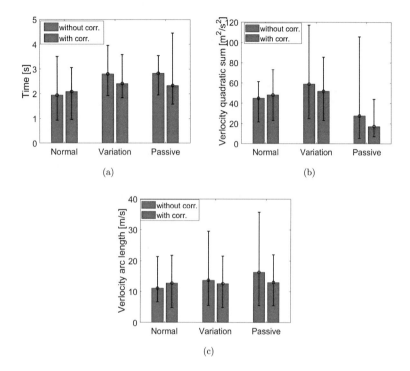

Figure 8.8: Comparison of performance metrics between two controllers (blue: without online correction, red: with online correction, lower values are better): (a) Task completion time for quantifying fluency, (b) Quadratic sum of human velocities for quantifying energy consumption, (c) Arc length of human velocity profile for quantifying movement smoothness.

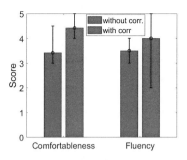

Figure 8.9: Subjective evaluation of the handover in comfortableness and fluency (blue: without online correction, red: with online correction, 1: unsatisfied, 5: very good, higher values are better).

in the normal mode, the values are higher (7.15% for velocity quadratic sum and 14.83% for the total arc length) .

Lastly, as shown in Fig. 8.9, the controller with online correction received a 29.27% higher score on confromableness than the one without online correction.

Discussion

The main outcome of the proposed method is improvement of success rate, comfortableness, fluency and human effort in a human-robot handover task relative to a non-adaptive controller, showing an ability to deal with dynamical and unexpected human motion.

The most important aspect is to adapt the spatial scaling of the nonlinear forcing term online based on the predicted handover location. The theoretic background is the invariant property of DMP [48]. Unlike previous research, the goal in this work is not to accurately predict the handover location but rather an early detection and fast response to the human motion trend. Hence, an online correction term is added using a simple gradient-type update law based on the human motion prediction error.

Although the final tracking performance of a DMP-based controller should theoretically not rely on the nonlinear term since this part always vanishes at the end of the trajectory. Our experiments show that in practice an improper spatial scaling of the non-linear term can also cause a failure in handovers. It drives the robot in the wrong direction, forces the human to adapt, and makes the error too large to compensate by the linear term. On the other hand, adding an online correction in the non-linear term leads to a significant

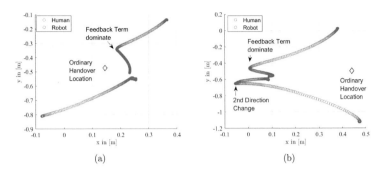

(a)   (b)

Figure 8.10: The paths of human and robot in two trials (blue : human red: robot): (a) without online correction, (b) with online correction

improvement in task performance.

Another important finding is that the proposed method enhances the legibility of the robot motion. According to the definition in [172], "legible motion enables an observer to quickly and confidently infer the correct goal". Hence, it plays an essential role in seamless human-robot cooperation. Fig. 8.10 shows the paths of human and robot in two trials. In both cases, the human changed its moving direction at the beginning of the approaching phase. It can be clearly seen in Fig. 8.10(a) that without correction, the robot moved firstly towards the pre-fixed ordinary handover location (in the middle of the line connecting the human and the robot), made its intention unclear to the human. As long as the feedback term dominated the motion planning, it drove the robot to move towards human, which caused a sharp turn in the path. Fig. 8.10(b) shows that the proposed controller achieved an early reaction of human motion change, and its path was more apparent to the actual handover location. Even the participant performed one more change of its moving direction, the robot still followed well. In summary, the advantage of the proposed method is that it adapts to the human partner's motion not only through the goal-attraction term (as the conventional DMP does) but also through the shape-attraction term that is learned from human demonstrations. Hence, the online adaption of robot motion is reactive, legible, and human-like, which significantly enhances the human comfortless.

However, it should be pointed out that the proposed method performed slightly worse than the non-adaptive controller in the normal handover mode, in which the goal position did not change a lot. In this case, the human motion prediction error might have been caused by other effects. Nevertheless, we still tried to adjust the spatial scaling to reduce the error. It caused an over-correction of the robot motion and made its behaviour unexpected for human.

This work still has many limitations. First of all, due to the small number of participants, the results can only indicate the potential impact of the proposed framework on performance enhancement in human-robot handover tasks. To draw more general conclusions, further tests with a large number of participants are needed.

Secondly, the orientation of the robot end-effector has not been considered. Instead of a fixed configuration, the orientation should be adjusted online based on the human trajectory, grasping type, and the grasped object. According to our observations in the experiments and the user feedback, online adaption of the end-effector orientation would further increase the handover fluency and reduce human effort.

Another limitation is that this work lacks discussion of the impact of the transferred object. A previous study in [173] shows that the object mass can influence the handover duration. Future studies should consider using objects with different sizes and weights, and analyze their influences on the duration, handover location, grasping configuration, etc.

8.3 Experimental study: object handling

This section presents the experimental study on the proposed framework with physical collaboration tasks and focuses on interaction control. Fig. 8.11 illustrates the fundamental components of a PHRC task, namely transporting a rigid object. As analyzed in Chapter 6, the task can be represented by a shared control structure, in which both the human and the robot are modeled as mechanical impedance and generate forces/torques to influence the object's dynamic. Within the concept of this work, both the reference trajectory and impedance parameters of the human are unknown. The task of the robot involves estimation of the human reference trajectory and learning its own optimal impedance parameters. For evaluation, two experiments were performed. The first one aimed to validate the Q-learning-based adaptive impedance control. The second one extended the control structure with human reference prediction.

Both experiments were performed on the *Franka Emika* robot. Note that the optical motion capture system was not used for this scenario. The human hand motion was computed based on the robot end-effector's position and velocity measurement. A direct measurement of the acceleration was not available. Usually it can be calculated by differentiating the velocity signal, which, however, can cause large noises and errors. In this work, a Kalman-filter with constant acceleration model (see Chapter 2) was

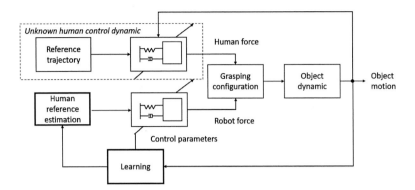

Figure 8.11: Overview of the shared control structure in PHRC

implemented to determine the acceleration. Furthermore, the robot's control interface provides an estimated value of the external force acted on the end-effector.

8.3.1 Validation of the Q-learning based adaptive impedance control

For simplification, the experiment only contains a 2D translation motion. As shown in Fig. 8.12, the human operator and the robot collaboratively transferred an object along a prescribed path between two points, A and B. The robot end-effector's trajectory is pre-defined by a quintic polynomial as a function of time by taking A and B as initial- and end position. During the execution, the human made a corrective motion to bring the object to point C, which is unknown to the robot (red dashed line in Fig. 8.12). As long as the correction was finished, the robot should regenerate a new point-to-point motion from C to B that drives it back to the original path (blue dashed line in Fig. 8.12). If the proposed control algorithm works appropriately, the robot will decrease its impedance when the human tries to dominate the collaboration. On the contrary, if the human has no intention of motion correction, the robot should maintain a high impedance to reduce the tracking error.

Results

Fig. 8.13 shows the experiment results. Only the values in y-dimension are plotted since in x-dimension the results are similar. The gray zone describes the phase that human moved the object to position C. At first (0-15 s) the robot maintained a high stiffness

Figure 8.12: Experiment setup, from left to right: 1) sketch of the task, 2) starting position A, 3) end position B, 4) human desired new position C

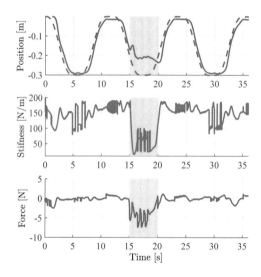

Figure 8.13: Experiment results in y-dimension: (from top to bottom) 1) reference (dash line) and actual position (solid line), 2) adaption of robot stiffness, 3) estimated interaction force.

in order to reduce the tracking error. Meanwhile the human operator did not generate large force, which means he was generally agree with the robot control behavior.

In the gray zone (15-20 s), the human tried to dominate the motion by increasing the force. At the same time robot decreased its stiffness to reduce the interaction force so that the human can easily move the robot with a small force (≈ 5 N). Note that one can observe light oscillation in both force and stiffness signal in the human-dominate phase. The main reason is that during this phase, from the robot's point of view the motion error was large, hence it tried to increase its stiffness inquiringly to bring the object back to the prescribed trajectory. As long as human disagreed with that it decreased the stiffness again. Since the stiffness adaption took place every 10 ms (running period of the learning algorithm). the signals appear to be noisy.

In the last phase (20-36 s), human reduced the force since the corrective motion was finished. It can be seen that the robot increased its stiffness and brought the object back to the original trajectory.

The results above show that the proposed method is able to continuously adapt the impedance control parameter according to task performance and human inputs. On the other hand, since human reference estimation was not included in this experiment, during the human correction phase, the robot still generated a force to drag the object back to the original path. The performance can be further improved by combing the adaptive impedance control with human reference estimation so that the robot can proactively adjust its motion reference.

8.3.2 Human reference estimation

As discussed in Chapter 6, in PHRC, since the human and the robot usually need to maintain a task-specified group formation, their dynamics are tightly coupled by kinematic constraints. Then human motion is more predictable since the uncertainties become smaller. In this work, human reference is estimated based on the coupled dynamics of the human and the robot.

Considering the coupled mass-spring-damper system in Figure 8.14), x_o represents the object's position, $x_{H,d}, x_{R,d}$ represents the reference positions of human and robot respectively, $M = M_o + M_d$ is the sum of the object and robot equivalent mass. The coupled dynamic is described as follows:

$$K_H(x_{H,d} - x_o) - D_H\dot{x}_o = -M\ddot{x}_o - D_d\dot{x}_o + K_d(x_{R,d} - x_o) \qquad (8.6)$$

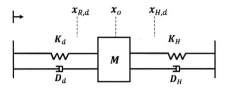

Figure 8.14: Second-order system formed by human and impedance controlled robot

Then the human reference position $\boldsymbol{x}_{H,d}$ becomes:

$$\boldsymbol{x}_{H,d} = \boldsymbol{K}_H^{-1}\left(-\boldsymbol{M}\ddot{\boldsymbol{x}} + (\boldsymbol{D}_H - \boldsymbol{D}_d)\dot{\boldsymbol{x}} + (\boldsymbol{K}_H - \boldsymbol{K}_d)\boldsymbol{x} + \boldsymbol{K}_d\boldsymbol{x}_{R,d}\right), \qquad (8.7)$$

which can be formulated as a linear equation:

$$\boldsymbol{x}_{H,d} = \boldsymbol{\Psi}\boldsymbol{\beta}, \qquad (8.8)$$

where

$$\boldsymbol{\Psi} = \begin{pmatrix} \boldsymbol{I} & \boldsymbol{x}_o & \dot{\boldsymbol{x}}_o & \ddot{\boldsymbol{x}}_o \end{pmatrix}, \quad \boldsymbol{\beta} = \begin{pmatrix} \boldsymbol{K}_H^{-1}\boldsymbol{K}_d\boldsymbol{x}_{R,d} \\ \boldsymbol{K}_h^{-1}(\boldsymbol{D}_H - \boldsymbol{D}_d) \\ \boldsymbol{K}_H^{-1}(\boldsymbol{K}_H - \boldsymbol{K}_d) \\ -\boldsymbol{M} \end{pmatrix} \qquad (8.9)$$

The parameter set $\boldsymbol{\beta}$ can be identified using least-squares method, such that:

$$\hat{\boldsymbol{\beta}} = \left(\boldsymbol{\Psi}^T\boldsymbol{\Psi}\right)^{-1}\boldsymbol{\Psi}^T\boldsymbol{x}_{H,d}. \qquad (8.10)$$

Note that Eq. (8.10) only requires measurement of motion data, and no force information is needed.

However, one problem is that in reality, the human reference $\boldsymbol{x}_{H,d}$ is a latent variable and cannot be directly measured. To deal with this issue, it is assumed that the future position of the human at the next sampling equals to the current reference position, i.e., $\boldsymbol{x}_{H,d}(k) = \boldsymbol{x}_H(k+1)$, which means the human motor control is always able to compensate the motion error within one sampling period. Apparently, it might not be completely true in reality. However, the error should not be extremely large, especially within a small time range.

Based on the identified parameter set $\hat{\boldsymbol{\beta}}$ and the new incoming measurement $\boldsymbol{\Psi}^\star$, the estimated human reference is computed as:

$$\hat{\boldsymbol{x}}_{H,d} = \boldsymbol{\Psi}^\star\hat{\boldsymbol{\beta}}. \qquad (8.11)$$

Algorithm

As in all, the proposed adaptive impedance control based on Q-learning with human reference prediction is summarized in Algorithm 2.

Algorithm 2 Q-Learning impedance control

Require: measurement of motion $x_R(k)$, interaction force $f(k)$

   ▷ *Initialization*

1: Define an admissible control input u_0

2: **if** 10 measurements collected **then** Calculate parameter set $\hat{\beta}_0$ using Eq. (8.10)

3: **end if**

   ▷ *Reference update*

4: Update parameter vector $\hat{\beta}$ using RLS

5: Calculate estimated human reference $\hat{x}_{h,d}$ using Eq. (8.11)

6: Calculate robot reference x_d based on $\hat{x}_{H,d}$.

   ▷ *Q- learning*

7: Calculate $r(k)$ and $\Phi(k)$ using Eq. (7.8),(7.28)

8: **while** $\|S^+ - S\| \geq \varepsilon$ **do**

9:    $S^+ \leftarrow S - l \otimes \left(\Phi(k)\left(S^T\Phi(k) - r(k)\right)\right)$

10: **end while**

11: Calculate robot impedance parameters based on S

12: Adjust the impedance parameters using Eq. (7.37),(7.38) based on $f(k)$

13: Calculate the control input $\tau(k+1)$ in Eq. (7.35)

8.3.3 Evaluation of the whole framework

In the experiment, the human operator held the robot's end-effector (Figure 8.15(a)) and followed a 2D reference path (Figure 8.15(b)) along the pre-specified direction, which is marked by the black arrows. Firstly the human operator was asked to track the inner "eight" curve (blue). Then to further test the flexibility of the proposed control algorithm, the human was asked to switch to the outer circle path (red) when the marked starting point was reached for the second time. Functions of both paths are shown as follows:

$$\text{outer path:} \begin{cases} x = 0.133\sin(2\pi i), \\ y = 0.133\cos(2\pi i), \end{cases}$$

$$\text{inner path:} \begin{cases} x = 0.133\sin(2\pi i)\cos(2\pi i), \\ y = 0.133\sin(2\pi i), \end{cases} \tag{8.12}$$

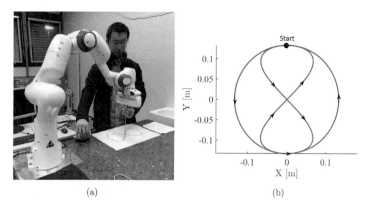

(a)   (b)

Figure 8.15: a) picture of the experimental setup with a *Franka Emika* robot, b) reference
path.

where $i \in [0, 1]$.

Note that the reference path was not given to the robot, except the starting point.
As long as the human tried to move the end-effector, the robot started to estimate
the human reference position and used it as its own reference for the inner adaptive
impedance control loop. The initial parameter sets of the Q-learning algorithm are
given as follows:

$$\mathbf{Q}_R = \begin{pmatrix} 2 & 0 & 0 & 0 \\ 0 & 2 & 0 & 0 \\ 0 & 0 & 2 & 0 \\ 0 & 0 & 0 & 2 \end{pmatrix}, \mathbf{R}_R = \begin{pmatrix} 0.02 & 0 \\ 0 & 0.02 \end{pmatrix}$$

$$\boldsymbol{S}_{xx} = \boldsymbol{I}^{4 \times 4}, \boldsymbol{S}_{uu} = \begin{pmatrix} 0.1 & 0 \\ 0 & 0.1 \end{pmatrix}, \boldsymbol{S}_{xu} = \begin{pmatrix} 10 & 0 & 2.8 & 0 \\ 0 & 10 & 0 & 2.8 \end{pmatrix}. \tag{8.13}$$

Totally 8 participants joined in the experimental study. For comparison, three different
impedance control setups were tested. Besides the proposed method, a conventional
impedance controller with low stiffness and damping was used. In this case, the robot
behaved compliantly so that the human would dominate the collaboration. Then an
adaptive impedance controller with human reference estimation was tested. The difference is that only the position reference was adapted, while the stiffness and damping

were kept constant and high for better tracking. The constant impedance parameter sets used in the other two setups are given as follows:

$$K_{d,low} = \begin{pmatrix} 10 & 0 \\ 0 & 10 \end{pmatrix}, \ D_{d,low} = \begin{pmatrix} 6.32 & 0 \\ 0 & 6.32 \end{pmatrix}, \tag{8.14}$$

$$K_{d,high} = \begin{pmatrix} 100 & 0 \\ 0 & 100 \end{pmatrix}, \ D_{d,high} = \begin{pmatrix} 20 & 0 \\ 0 & 20 \end{pmatrix}. \tag{8.15}$$

All participants repeated the task three times for each control setup. The total error of the path following, total execution time, and total force produced by the human operator are chosen as performance indexes for evaluation. Moreover, user experience is also considered. After performing all the measurements, the participants were asked the following two questions:

Q1 *Which control setup provides the best comfortableness?*

Q2 *Which control setup provides the best assistance?*

Results

Fig. 8.16 shows the average and variance of all the performance indexes as mentioned above. In Fig. 8.16(a), the integrated tracking error was approximately determined by calculating the area of the reference- and actual path. It can be seen that impedance control with human reference estimation (for both constant and variable impedance parameters) significantly reduced the tracking error. By taking a further look at Fig. 8.17(a), especially when the human operator performed the outer circular path, significant error can be observed. In the other two setups, since the robot was able to generate assistive force based on the estimation of human's motion reference, the tracking performance became much better (Fig. 8.17(c) and Fig. 8.17(b)). Results in these setups are close to each other, showing that the impedance adaption does not contribute much to error reduction.

By comparing the execution time of one single demonstration (Fig. 8.16(b)), one can see that the proposed method provided the highest time efficiency.

The integral of the interaction force over time can be regarded as a measurement of total human effort. Fig. 8.16(c) indicates that the human effort is proportional to the robot's impedance, i.e., high impedance > variable impedance > low impedance. The result does not rely on whether the human reference prediction is provided.

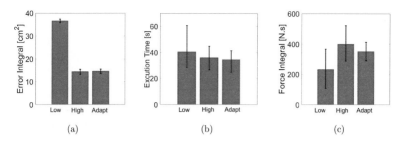

(a) (b) (c)

Figure 8.16: Multiple comparison of the performance indexes between all three setups: (a) integrated path tracking error, (b) execution time of one demonstration, (c) intergrated contact force.

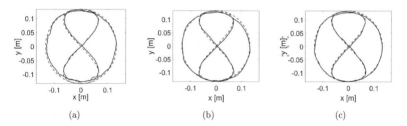

(a) (b) (c)

Figure 8.17: Path tracking performance of all three setups (average values): (a) standard impedance control, (b) impedance control with human reference estimation, (c) proposed method. The dashed- and solid line represent the reference- and real path respectively.

Fig. 8.18 plots the force and position in x-direction over time in one single demonstration. In the low impedance case, a large peak of contact force is observed when the human started to follow the outer circular path (25 s - 30 s). In the other two setups with human reference estimation, this peak value is significantly reduced, indicating that the robot recognized the change of human trajectory and provides assistant force to reduce human error. Nevertheless, large force peaks are observed when the human intended to change the moving directions. One possible reason is that the robot's reaction was not fast enough as the human expected. Therefore, the human intended to make correction. Increasing the prediction horizons of the human trajectory can be a solution of this problem. Another possible reason is due to the coupling effect in robot impedance control, which generates unexpected coupling forces in different directions.

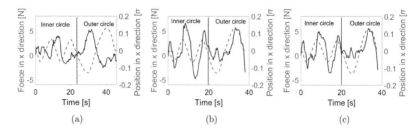

(a)   (b)   (c)

Figure 8.18: Time History of the interaction force and position in x-direction from one single demonstration: (a) standard impedance control, (b) impedance control with human reference estimation, (c) proposed method. The dashed- and solid line represent the position- and force respectively.

	Best comfortableness	Best assistance
Setup 1	1	1
Setup 2	2	3
Setup 3	5	4

Table 8.2: Statistic of the user experience. Setup 1: standard impedance control with low stiffness and damping, Setup 2: impedance control with human reference estimation and constant parameters, Setup 3: proposed method.

Additionally, a survey of user experience about all three setups is summarized in Tab. 8.2. 5 of 8 participants chose the proposed method (setup 3) as "the most comfortable one" since the robot was "light" and can provide "active contribution" as well. Regarding "the best assistance", 4 of 8 chose the proposed method since they thought the robot "jointly moved with human" and "ran smoothly". On the other hand, 3 of 8 chose the setup with a constant impedance parameter set. The participants explained that due to the impedance adaption, sometimes the robot became "inconsistent" and "hard to control". It is mainly due to the dynamical change of the impedance parameters (see Figure 8.19). The study indicates that a human user might have its own initiative of when and how frequently the robot should adapt its control behaviors.

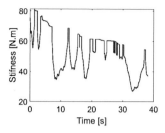

Figure 8.19: Adaption of the robot stiffness in x-direction performed by the proposed method from one single demonstration

Discussion

The most important outcome of this study is that the robot making a proactive contribution improves the task performance significantly and yields a more accurate and efficient PHRC. A reliable and fast estimation of the human's intention is a prerequisite. Furthermore, the learning-based adaption of the impedance control parameters enhances the user experience, especially in comfortableness.

From the role allocation point of view, the experiment indicates that the human-dominated strategy is not optimal. Under this concept, the human needs to take care of task execution and coordinate with the robot simultaneously, which strongly increases its workload (both physical and psychological). Hence, many recent research works suggest a dynamic role allocation strategy, in which the human and the robot continuously adjust their roles/dominance during collaboration. In PHRC, it can be achieved by adapting the load distribution [165], the cost function [166] or the impedance parameters (this work). However, according to the above user study, several human users do not favor this adaption mechanism. Similar conclusions have been drawn in one related work [174]. First of all, "people collaborate best with a proactive robot, yielding better team fluency and high subjective ratings." However, "they prefer having control of when the robot should help". In other words, even sometimes humans are willing to give up their dominance in collaboration, they still have their own role allocation rule, and the robot should observe.

This work still has many limitations. As a proof of concept, only two DOFs are considered. Further experiments on rotational DOFs are required. In this case, the system model becomes nonlinear. Although the Q-learning-based approach can also work for nonlinear systems [141], validations are still needed. Secondly, no constraint has been included in the design, which is critical to safety-related applications. Last but not least, this design considered human motion prediction in a simple manner (linear model + one-step prediction). Uncertainty-aware long-term prediction should be studied in future works, which can further enhance the task performance.

8.4 Summary

This chapter presented the experimental study of the proposed learning and control framework. Motivated by the special feature of impedance control, i.e., it can work for both motion and contact force regulation, the design aims to extend the conventional impedance control framework with a learning module so that the robot can adapt to as many types of collaboration tasks as possible. The learning module consists of human motion prediction, robot trajectory planning, and control parameters optimization. The functionalities of each component, as well as their combinations, are validated by different experiments.

Totally four experiments were performed. Two of them were designed for human-robot handovers, which mainly focus on online robot trajectory generation. The difficulty lies in that the human-desired handover location is unknown. The first design is a data-driven approach, in which the human hand position is predicted via online GP regression. The outputs are used for generating a smooth point-to-point motion for tracking the human hand position. To ensure safety, the stiffness matrix of the impedance controller is adjusted through a danger index that defined based on prediction uncertainties. Experiment results show that the basic functionalities are achieved. However, the method is sensitive to human motion variation and lacks flexibility.

To overcome this problem, the second design combines GP regression with DMP to fit its nonlinear shape attraction term. This design yields a hybrid representation of human motion. On the one hand, the linear part of DMP provides a physical interpretation of the "attractive" behavior in handovers. On the other hand, the shape attraction term in DMP is learned through a data-driven approach. Moreover, the invariant properties of DMP enable generating various trajectories by changing only a small number of parameters, which significantly reduces the computational load. Experimental results show that this approach enhances the success rate of handovers, reduces the total execution time, and improves fluency and human comfortableness.

The last two experiments were designed for collaborative object handling and transporting, focusing on interaction control. Firstly the Q-learning-based adaptive impedance control was validated. The results show that the robot decreased impedance when the human intended to correct its motion and maintained high impedance to reduce the tracking error when no human intervention occurred. Then to increase the flexibility, the control framework was extended with a human reference estimator, which provided a short-term human motion prediction via linear regression. Results show that the proposed framework significantly reduces the motion error and enhances human comfortableness. However, the user study indicates that not all the participants prefer a dynamically adapted impedance behavior. It would be interesting to study further when and how the adaption should be performed during the interaction with a human.

Many possible combinations have not been included in this chapter and can be considered in future work. The first one is to combine reinforcement learning with GP, which employs statistical generative models for value functions [175]. Compared to the linear value function approximation methods used in this work, the GP-based representation can extend the generalizability of the value function-based learning method, particularly for nonlinear systems. Another possible approach is to integrate the concept of DMP into impedance control, which yields impedance-based primitives [176] so that the task execution skills can be directly represented and learned from the interaction control point of view.

9 Conclusion and Outlook

9.1 Conclusion

Researches in recent years have achieved great success in the guarantee of safety in human-robot interaction, yielding a new generation of robots, namely the "cobots", which can work with humans in close proximity. However, due to the lack of ability to understand and coordinate with humans, the "co" in most cobots still refers to "coexistence" rather than "collaboration". The thesis proposed a development concept of an adaptive control framework with a novel physical and data-driven approach towards a real collaborative robot.

Online prediction of human motion trajectories

Modeling and predicting human movements in considerations of both accuracy and computational efficiency are challenging. Instead of precise but labor-intensive biomechanical models, this thesis aims to find a modeling approach that can cover the most critical human motion features, require little sensor information and adapt online. For this purpose, the thesis began with well-developed recursive Bayesian estimation approaches with simple linear physical kinematic models. Two concepts, Kalman filter with constant acceleration model and recursive least square filter with minimum jerk model, were tested on a human reaching motion data set. These methods have simple mathematical formulations and low implementation complexities. Results indicate that they are suitable for short-term prediction with fewer uncertainties in a static environment. However, the error grows in proportion to prediction horizons, showing their limitation in long-term predictions.

It can be seen that even for a simple reaching movement with consideration of only translational degrees of freedom, linear dynamic models with additive Gaussian noises are only valid in a small range around a certain velocity. As the next development step, the thesis investigated data-driven modeling approaches supported by machine learning techniques. Motivated by the findings from human motor control research, i.e., the human central nervous system works as an optimal feedback control system, the first study focused on identifying the objective function that the human tries to optimize. The state-of-the-art inverse optimal control based on the principle of maximum entropy was implemented. The method requires solving two constrained optimization problems: one for the determination of the cost function and the other for the reproduction of human trajectories. Hence, the implementation complexity is high. Results show that this approach enhances the prediction performance for long look-ahead time but not significantly. The online adaptability is low since the method usually requires complete human trajectory observations.

In all the above methods, one needs to restrict the class of functions that are used to model the input and output relations (e.g., linear functions). If the chosen class of functions cannot model the system well, the prediction will be inaccurate. To overcome this problem, the thesis turned the attention to non-parametric approaches. A recursive sparse Gaussian process regression algorithm was applied. Comparing to standard GP, this method reduces the computational complexity and makes the model adaptive to incoming data. Results show that this method performed well in both short-term and long-term predictions. Moreover, the recursive algorithm is able to deal with new data from unknown batches that are not included in the training set. Nevertheless, it still requires several iterations to learn the distribution of new data. During this process, the predictions are not reliable.

After investigating both physical and data-driven approaches, it is concluded that their properties are somehow complementary. Simple physical models usually generalize well in their valid ranges, do not require training data, but can only represent a limited part of the system behavior. Data-driven approaches have a strong ability to represent complex systems, are uncertainty-aware, but can only work in the regime in which data is available. Therefore, a hybrid physical and data-driven approach seems very promising. Dynamic movement primitives are among the most extensively studied hybrid modeling approaches, extending a physically well-understood linear attractor dynamic with a learnable nonlinear forcing term. This thesis extended the conventional DMP formulation with a GP-based shape-attraction term to (1) utilize its ability to describe nonlinear dynamics, (2) reduce the manual parameter tuning effort. Furthermore, to solve several practical issues, especially the sensibility of the DMP model to the varying goal position, a weighting factor to adjust the dominance of linear and nonlinear terms and a rotation-based spatial scaling factor was designed. Results show that this approach outperforms all the other methods that have been implemented in this work.

In summary, the outcomes of this thesis indicate that hybrid approaches with combinations of physically well-understood models and data-driven modulations would be a promising concept of learning and predicting human motions.

Adaptive robot interaction control

Human motion prediction helps to generate human-aware robot trajectories. The thesis's second part focuses on the execution phase. If no physical contact occurs during collaboration, only motion control of the robot needs to be considered. However, many collaboration tasks require physical interactions between humans and robots, which are more challenging to control design. Besides safety, robots should be able to make proactive contributions to reduce humans' physical and coordination effort further.

For this purpose, the thesis first studied the group dynamic behaviors when humans and robots jointly move a rigid object. Instead of a comprehensive kinematic and dynamic modeling of each agent with consideration of all degrees of freedoms, a more efficient approach is to incorporate the compliance control behavior of both robots and humans into the object dynamic through physical constraints. Impedance control is a well-known compliant control approach in which both humans and robots can be regarded as a second-order mass-spring-damper system. Hence, from the modeling point of view, physical human-robot collaboration tasks can be approximately described by a serious of couple mechanical impedance, which significantly simplifies the model complexity.

The coordination between humans and robots can be formulated through differential games, in which all the participants aim to solve an optimization problem cooperatively. Note that even for the simplified model, it is difficult to design a control algorithm using an explicit model-based approach since the human control parameters are unknown and time-varying. The thesis again combined the machine learning techniques with physically well-studied control methods to solve this problem, yields an adaptive impedance control framework based on reinforcement learning. The key idea is to adapt the control parameters using the Q-learning algorithm during the interaction with the human. Furthermore, concepts of contact force and constraints handling were also discussed.

Experimental validation

The proposed learning and control framework has been validated by several human-robot collaboration tasks on a 7-DOF robot platform. Its low-level control interface provides rigid body dynamic parameters, forward/backward kinematics, and joint parameter (including joint torque) measurements. The proposed framework was programmed on an external PC and communicated with the robot through Ethernet. The human motion was recorded by an optical motion capture system. The experiments contained two typical benchmark applications, namely object-handover and object-handling, so both contact-free and physical interactions were included.

Results proved that the proposed framework is able to:

- learn human motion models with small amount of training data,

- perform reliable online human motion prediction and an early reaction to human motion variation,

- provide proactive contributions on physical collaborations and behave compliantly in reaction to human forces,

- enhance both task performance (success rate, execution time, position error) and user experience (comfortableness, fluency).

9.2 Outlook

The outlook is presented from two perspectives: (a) practical extensions that help the proposed framework go beyond the laboratory setups and work in real-world applications; (b) theoretical extensions to answer some fundamental questions of the hybrid physical and data-driven approach for both modeling and control.

Practical extensions

All practical designs in this work aim to provide methodological validations of the proposed concept. Hence, many simplifications have been made, limiting the proposed framework's usability out of the lab. In this sense, the following practical extensions are suggested:

- Considering rotational DOFs in human motion and impedance adaption:
  All the experiments in this thesis only studied translational motions. Many tasks, such as collaborative assembling work, packaging, inspection, etc., involve rotational movements. Even for the two scenarios presented in this work, extending the current setup by considering rotational DOFs could enhance the flexibility and human comfortableness. Moreover, rotational compliance also plays an essential role in interaction control and should be investigated.

- Considering simplified human body kinematics:
  In this thesis, the human hand was regarded as a point mass. In order to learn more complex human skills and refine the robot trajectory planning in a shared

workspace with human, geometric and kinematic characteristics of several human body parts (e.g., arm) needs to be considered. Note that the extension should not dramatically increase the model's complexity but rather in a simplified manner, e.g., a skeleton model or a series of rigid bodies with regular shapes.

- Considering multi-modal sensory information and control interfaces:
  In this thesis, the human motions were perceived by a limited type and number of the sensors, which obtained only partial information of the human activities. With the development of sensor technology, more and more light-weight, cost-effective and reliable wearable sensors have come on the market. Multi-modal and multi-dimensional information enables robot to address complex interaction scenarios and improve the task execution performances. For instance, in the handover task, installing extra force sensors on the robot gripper can provide feedback if the grasping is successful. Nevertheless, a fair compromises between completeness and complexity should be made.

Theoretical extensions

Although this thesis presents a large potential of hybrid physical-based and data-driven approaches in both human modeling and robot control design, many open questions remain. In Chapter 5, several possible future trends in human motion modeling and prediction were proposed and will not be repeated here. Furthermore, some other interesting research questions are listed as follows:

- Which physical model should be used, and how should it be combined with a data-driven model?
  The hybrid modeling approach in this work was developed based on Dynamic movement primitives, in which the physical description is provided by a second-order mechanical model. This model type is particularly suitable for a goal-oriented reaching movement. Besides, many other types of physical models, e.g., fluid dynamics, energy fields, can also be used to describe human motion. On the other hand, it is worth deeply investigating how to combine physical models with data-driven approaches. For instance, in DMP, they are combined through an additive relationship since both are regarded as forces so that their sum will generate human motion. Of course, it is not the only concept. For example, the combination can also be achieved by a probabilistic description, in which the physical model is used as kernel functions.

- How to achieve constrained reinforcement learning with partial information of the environment provided by simplified physical models?

Reinforcement learning is a model-free approach, and the optimal policy is learned through interacting with the environment. However, there is no guarantee that the interaction process will not generate any safety-critical state. Hence, constrained reinforcement learning is an urgent research topic. As discussed in Chapter 7, constraints handling without physical knowledge of the system is unachievable. A possible solution would be using a reduced-size physical model to describe the most relevant part of the system to constraints handling. Before applying to hardware, the learned policy can be simulated in this reduced model to check if it will violate the constraints. This step can be regarded as a "supervision" of the learning process. The unsafe policy can be labeled and avoided in future learning iterations.

- How to bridge the gap between different research communities?
  Human-robot collaboration is an interdisciplinary research area compromising various subjects. In recent years, several pioneering works have successfully associated classical robotics with biomechanics, human-computer iteration, and artificial intelligence. This thesis suggests paying more attention to cognitive sciences and psychology. The former helps to understand better how humans process information, represent knowledge and make decisions. Such skills can be transferred into robots. The latter plays an essential role in designing robot behaviors and coordination strategies that are more natural and acceptable to humans.

Appendices

A Inverse Optimal Control with Local Optimality

Recalling the principle of maximum entropy, the probability of the optimal control input and state trajectories id proportional to the exponential of the cost uncounted along the whole trajectory, namely:

$$p\left(\boldsymbol{\zeta}^*|\mathbf{x}_0\right) = \frac{\exp\left(-J\left(\boldsymbol{\zeta}^*\right)\right)}{\int \exp\left(-J\left(\boldsymbol{\zeta}\right)\right)\mathrm{d}\boldsymbol{\zeta}}. \tag{A.1}$$

Computing the integral in the denominator is non-trial. Hence, the Laplace Approximation is applied, which models the distribution of $\boldsymbol{\zeta}$ as Gaussian. Moreover, under the assumption that human performs a local optimization rather than a global solution, the probability can be simplified by taking a second-order Taylor expansion of J centered on $\boldsymbol{\zeta}^*$. It yields:

$$J\left(\boldsymbol{\zeta}\right) \approx J(\boldsymbol{\zeta}^*) + \left(\boldsymbol{\zeta} - \boldsymbol{\zeta}^*\right)^{\mathrm{T}}\mathbf{g} + \frac{1}{2}\left(\boldsymbol{\zeta} - \boldsymbol{\zeta}^*\right)^T\mathbf{H}\left(\boldsymbol{\zeta} - \boldsymbol{\zeta}^*\right), \tag{A.2}$$

where $\mathbf{g} = \frac{\partial J}{\partial \boldsymbol{\zeta}^*}$ and $\mathbf{H} = \frac{\partial^2 J}{\partial \boldsymbol{\zeta}^{*2}}$. Taking the exponential:

$$\exp\left(-J\left(\boldsymbol{\zeta}\right)\right) \approx \exp\left(-J\left(\boldsymbol{\zeta}^*\right)\right)\exp\left(-\left(\boldsymbol{\zeta} - \boldsymbol{\zeta}^*\right)^T\mathbf{g}\right)\exp\left(-\frac{1}{2}\left(\boldsymbol{\zeta} - \boldsymbol{\zeta}^*\right)^T\mathbf{H}\left(\boldsymbol{\zeta} - \boldsymbol{\zeta}^*\right)\right). \tag{A.3}$$

Then

$$p\left(\boldsymbol{\zeta}^*|\boldsymbol{\theta}, \mathbf{x}_0\right) = \frac{\exp\left(-J(\boldsymbol{\zeta}^*)\right)}{\int \exp\left(-J\left(\boldsymbol{\zeta}^*\right)\right)\exp\left(-\left(\boldsymbol{\zeta} - \boldsymbol{\zeta}^*\right)^T\mathbf{g}\right)\exp\left(-\frac{1}{2}\left(\boldsymbol{\zeta} - \boldsymbol{\zeta}^*\right)^T\mathbf{H}\left(\boldsymbol{\zeta} - \boldsymbol{\zeta}^*\right)\right)\mathrm{d}\boldsymbol{\zeta}} \tag{A.4}$$

$$= \left(\int \exp\left(-\left(\tilde{\boldsymbol{\zeta}} - \boldsymbol{\zeta}\right)^T\mathbf{g} - \frac{1}{2}\left(\tilde{\boldsymbol{\zeta}} - \boldsymbol{\zeta}\right)^T\mathbf{H}\left(\tilde{\boldsymbol{\zeta}} - \boldsymbol{\zeta}\right)\right)\mathrm{d}\tilde{\boldsymbol{\zeta}}\right)^{-1} \tag{A.5}$$

By making use of the standard result for the normalization of a Gaussian, the distribution $p\left(\boldsymbol{\zeta}^*|\boldsymbol{\theta}, \mathbf{x}_0\right)$ is obtained:

$$p\left(\boldsymbol{\zeta}^*|\boldsymbol{\theta}, \mathbf{x}_0\right) = \left(\exp\left(\frac{1}{2}\mathbf{g}^T\mathbf{H}^{-1}\mathbf{g}\right)\sqrt{\frac{(2\pi)^{mK}}{\det(\mathbf{H})}}\right)^{-1} \tag{A.6}$$

$$= \exp\left(-\frac{1}{2}\mathbf{g}^T\mathbf{H}^{-1}\mathbf{g}\right)\det(\mathbf{H})^{\frac{1}{2}}(2\pi)^{-\frac{mK}{2}}, \tag{A.7}$$

where

$$\mathbf{g} = \frac{\partial J}{\partial \mathbf{u}^*} + \frac{\partial \mathbf{x}^*}{\partial \mathbf{u}^*}^T \frac{\partial J}{\partial \mathbf{x}^*}, \tag{A.8}$$

$$\mathbf{H} = \frac{\partial^2 J}{\partial \mathbf{u}^{*2}} + \frac{\partial \mathbf{x}^*}{\partial \mathbf{u}^*}^T \frac{\partial^2 J}{\partial \mathbf{x}^{*2}} \frac{\partial \mathbf{x}^*}{\partial \mathbf{u}^*}. \tag{A.9}$$

Considering the following linear time-invariant system

$$\mathbf{x}\left(k+1\right) = \mathbf{A}\mathbf{x}\left(k\right) + \mathbf{B}\mathbf{u}\left(k\right), \quad \mathbf{x} \in \mathbb{R}^{n\times1}, \mathbf{u} \in \mathbb{R}^{m\times1}, \tag{A.10}$$

and quadratic feature functions:

$$\boldsymbol{\phi}\left(k\right) = \begin{pmatrix} \mathbf{x}\left(k\right) \otimes \mathbf{x}\left(k\right) \\ \mathbf{u}\left(k\right) \otimes \mathbf{u}\left(k\right) \end{pmatrix} \in \mathbb{R}^{(n+m)\times1}. \tag{A.11}$$

If an expert trajectory $\boldsymbol{\zeta}^* = (\mathbf{x}^*, \mathbf{u}^*)$ with N samples is available, the elements in $\mathbf{g}$ and $\mathbf{H}$ are determined as follows:

$$\frac{\partial J}{\partial \mathbf{u}^*} = \begin{pmatrix} \boldsymbol{\theta}_u \mathbf{u}(1) \\ \vdots \\ \boldsymbol{\theta}_u \mathbf{u}(K) \end{pmatrix}, \quad \frac{\partial^2 J}{\partial \mathbf{u}^{*2}} = \begin{pmatrix} \boldsymbol{\theta}_u & & \\ & \ddots & \\ & & \boldsymbol{\theta}_u \end{pmatrix}, \quad \boldsymbol{\theta}_u = \begin{pmatrix} \boldsymbol{\theta}(n+1) & & \\ & \ddots & \\ & & \boldsymbol{\theta}(n+m) \end{pmatrix}$$
$$\tag{A.12}$$

$$\frac{\partial J}{\partial \mathbf{x}^*} = \begin{pmatrix} \boldsymbol{\theta}_x \mathbf{x}(1) \\ \vdots \\ \boldsymbol{\theta}_x \mathbf{x}(K) \end{pmatrix}, \quad \frac{\partial^2 J}{\partial \mathbf{x}^{*2}} = \begin{pmatrix} \boldsymbol{\theta}_x & & \\ & \ddots & \\ & & \boldsymbol{\theta}_x \end{pmatrix}, \quad \boldsymbol{\theta}_x = \begin{pmatrix} \boldsymbol{\theta}(1) & & \\ & \ddots & \\ & & \boldsymbol{\theta}(n) \end{pmatrix} \tag{A.13}$$

$$\frac{\partial \mathbf{x}^*}{\partial \mathbf{u}^*}^T = \begin{pmatrix} \boldsymbol{D}_{1,1} & \cdots & \boldsymbol{D}_{1,K} \\ \vdots & \ddots & \vdots \\ \boldsymbol{D}_{K,1} & \cdots & \boldsymbol{D}_{K,K} \end{pmatrix}, \quad \boldsymbol{D}_{k_1,k_2} = \begin{cases} \mathbf{B}^T, & k_2 = k_1 + 1 \\ \boldsymbol{D}_{k_1,k_2-1}\mathbf{A}^T, & k_2 > k_1 + 1 \\ \mathbf{0}, & \text{else.} \end{cases} \tag{A.14}$$

Bibliography

[1] Arash Ajoudani, Andrea Maria Zanchettin, Serena Ivaldi, Alin Albu-Schäffer, Kazuhiro Kosuge, and Oussama Khatib. Progress and prospects of the human–robot collaboration. *Autonomous Robots*, 42(5):957–975, Jun 2018.

[2] Wilhelm Bauer, Manfred Bender, Martin Braun, Peter Rally, and Oliver Scholtz. Lightweight robots in manual assembly–best to start simply. *Frauenhofer-Institut für Arbeitswirtschaft und Organisation IAO, Stuttgart*, 2016.

[3] Luca Gualtieri, Erwin Rauch, and Renato Vidoni. Emerging research fields in safety and ergonomics in industrial collaborative robotics: A systematic literature review. *Robotics and Computer-Integrated Manufacturing*, 67:101998, 2021.

[4] Michael Flad. *Kooperative regelungskonzepte auf basis der spieltheorie und deren anwendung auf fahrerassistenzsysteme*, volume 2. KIT Scientific Publishing, 2017.

[5] Michael A Goodrich and Alan C Schultz. *Human-robot interaction: a survey*. Now Publishers Inc, 2008.

[6] Cordula Vesper, Ekaterina Abramova, Judith Bütepage, Francesca Ciardo, Benjamin Crossey, Alfred Effenberg, Dayana Hristova, April Karlinsky, Luke McEllin, Sari R. R. Nijssen, Laura Schmitz, and Basil Wahn. Joint action: Mental representations, shared information and general mechanisms for coordinating with others. *Frontiers in Psychology*, 7:2039, 2017.

[7] Cordula Vesper, Stephen Butterfill, Günther Knoblich, and Natalie Sebanz. A minimal architecture for joint action. *Neural Networks*, 23(8-9):998–1003, 2010.

[8] Jean-Michel Hoc. Towards a cognitive approach to human–machine cooperation in dynamic situations. *International journal of human-computer studies*, 54(4):509–540, 2001.

[9] Subhas Chandra Mukhopadhyay. Wearable sensors for human activity monitoring: A review. *IEEE sensors journal*, 15(3):1321–1330, 2014.

[10] Guilherme Maeda, Marco Ewerton, Gerhard Neumann, Rudolf Lioutikov, and Jan Peters. Phase estimation for fast action recognition and trajectory generation

in human-robot collaboration. *The International Journal of Robotics Research*, 36(13-14):1579–1594, 2017.

[11] H. Nemlekar, D. Dutia, and Z. Li. Object transfer point estimation for fluent human-robot handovers. In *2019 International Conference on Robotics and Automation (ICRA)*, pages 2627–2633, 2019.

[12] Valerio Ortenzi, Akansel Cosgun, Tommaso Pardi, Wesley P. Chan, Elizabeth Croft, and Dana Kulic. Object handovers: A review for robotics. *IEEE Transactions on Robotics*, pages 1–19, 2021.

[13] Shufei Li, Ruobing Wang, Pai Zheng, and Lihui Wang. Towards proactive human-robot collaboration: A foreseeable cognitive manufacturing paradigm. *Journal of Manufacturing Systems*, 60:547–552, 2021.

[14] Laura M Hiatt, Cody Narber, Esube Bekele, Sangeet S Khemlani, and J Gregory Trafton. Human modeling for human–robot collaboration. *The International Journal of Robotics Research*, 36(5-7):580–596, 2017.

[15] Andrey Rudenko, Luigi Palmieri, Michael Herman, Kris M Kitani, Dariu M Gavrila, and Kai O Arras. Human motion trajectory prediction: a survey. *The International Journal of Robotics Research*, 39(8):895–935, 2020.

[16] A. Elnagar. Prediction of moving objects in dynamic environments using kalman filters. In *Proceedings 2001 IEEE International Symposium on Computational Intelligence in Robotics and Automation (Cat. No.01EX515)*, pages 414–419, 2001.

[17] Andreas Møgelmose, Mohan M Trivedi, and Thomas B Moeslund. Trajectory analysis and prediction for improved pedestrian safety: Integrated framework and evaluations. In *2015 IEEE intelligent vehicles symposium (IV)*, pages 330–335. IEEE, 2015.

[18] Christoph Schöller, Vincent Aravantinos, Florian Lay, and Alois Knoll. What the constant velocity model can teach us about pedestrian motion prediction. *IEEE Robotics and Automation Letters*, 5(2):1696–1703, 2020.

[19] Sonia Duprey, Alexandre Naaim, Florent Moissenet, Mickaël Begon, and Laurence Chèze. Kinematic models of the upper limb joints for multibody kinematics optimisation: An overview. *Journal of Biomechanics*, 62:87–94, 2017. Human Movement Analysis: The Soft Tissue Artefact Issue.

[20] Muhammad Yahya, Jawad Ali Shah, Kushsairy Abdul Kadir, Zulkhairi M Yusof, Sheroz Khan, and Arif Warsi. Motion capture sensing techniques used in human upper limb motion: a review. *Sensor Review*, 2019.

[21] Markus Miezal, Bertram Taetz, and Gabriele Bleser. On inertial body tracking in the presence of model calibration errors. *Sensors*, 16(7), 2016.

[22] Gutemberg Guerra-Filho. Optical motion capture: Theory and implementation. *RITA*, 12(2):61–90, 2005.

[23] Cheng Xu, Jie He, Xiaotong Zhang, Cui Yao, and Po-Hsuan Tseng. Geometrical kinematic modeling on human motion using method of multi-sensor fusion. *Information Fusion*, 41:243–254, 2018.

[24] Yves Zimmermann, Alessandro Forino, Robert Riener, and Marco Hutter. Anyexo: A versatile and dynamic upper-limb rehabilitation robot. *IEEE Robotics and Automation Letters*, 4(4):3649–3656, 2019.

[25] Zhijun Li, Bo Huang, Zhifeng Ye, Mingdi Deng, and Chenguang Yang. Physical human–robot interaction of a robotic exoskeleton by admittance control. *IEEE Transactions on Industrial Electronics*, 65(12):9614–9624, 2018.

[26] Dragomir N Nenchev, Atsushi Konno, and Teppei Tsujita. *Humanoid robots: Modeling and control*. Butterworth-Heinemann, 2018.

[27] Sergey Jatsun, Andrei Malchikov, Oksana Loktionova, and Andrey Yatsun. Modeling of human-machine interaction in an industrial exoskeleton control system. In *International Conference on Interactive Collaborative Robotics*, pages 116–125. Springer, 2020.

[28] Morteza Asgari and Dustin L. Crouch. Estimating human upper limb impedance parameters from a state-of-the-art computational neuromusculoskeletal model. In *2021 43rd Annual International Conference of the IEEE Engineering in Medicine Biology Society (EMBC)*, pages 4820–4823, 2021.

[29] Luzheng Bi, Cuntai Guan, et al. A review on emg-based motor intention prediction of continuous human upper limb motion for human-robot collaboration. *Biomedical Signal Processing and Control*, 51:113–127, 2019.

[30] Tamar Flash and Neville Hogan. The coordination of arm movements: an experimentally confirmed mathematical model. *Journal of neuroscience*, 5(7):1688–1703, 1985.

[31] Y Uno, Mitsuo Kawato, and R Suzuki. Formation and control of optimal trajectory in human multijoint arm movement - minimum torque-change model. *Biological cybernetics*, 61:89–101, 02 1989.

[32] R. McN. Alexander. A minimum energy cost hypothesis for human arm trajecto-

ries. *Biological Cybernetics*, 76(2):97–105, 1997.

[33] Chiara Talignani Landi, Yujiao Cheng, Federica Ferraguti, Marcello Bonfè, Cristian Secchi, and Masayoshi Tomizuka. Prediction of human arm target for robot reaching movements. In *2019 IEEE/RSJ International Conference on Intelligent Robots and Systems (IROS)*, pages 5950–5957, 2019.

[34] Jing Zhao, Shiqiu Gong, Biyun Xie, Yaxing Duan, and Ziqiang Zhang. Human arm motion prediction in human-robot interaction based on a modified minimum jerk model. *Advanced Robotics*, 35(3-4):205–218, 2021.

[35] Z. Wang, M. Deisenroth, H. Ben Amor, D. Vogt, B. Schölkopf, and J. Peters. Probabilistic modeling of human movements for intention inference. In *Proceedings of Robotics: Science and Systems VIII*, page 8, 2012.

[36] Claudia Pérez-D'Arpino and Julie A. Shah. Fast target prediction of human reaching motion for cooperative human-robot manipulation tasks using time series classification. In *2015 IEEE International Conference on Robotics and Automation (ICRA)*, pages 6175–6182, 2015.

[37] Guoliang Fan, Xin Zhang, and Meng Ding. Gaussian process for human motion modeling: A comparative study. In *2011 IEEE International Workshop on Machine Learning for Signal Processing*, pages 1–6, 2011.

[38] Hongyi Liu and Lihui Wang. Human motion prediction for human-robot collaboration. *Journal of Manufacturing Systems*, 44:287–294, 2017. Special Issue on Latest advancements in manufacturing systems at NAMRC 45.

[39] Julieta Martinez, Michael J. Black, and Javier Romero. On human motion prediction using recurrent neural networks. In *2017 IEEE Conference on Computer Vision and Pattern Recognition (CVPR)*, pages 4674–4683, 2017.

[40] Yujiao Cheng, Weiye Zhao, Changliu Liu, and Masayoshi Tomizuka. Human motion prediction using semi-adaptable neural networks. In *2019 American Control Conference (ACC)*, pages 4884–4890. IEEE, 2019.

[41] Judith Bütepage, Michael J. Black, Danica Kragic, and Hedvig Kjellström. Deep representation learning for human motion prediction and classification. In *2017 IEEE Conference on Computer Vision and Pattern Recognition (CVPR)*, pages 1591–1599, 2017.

[42] Jim Mainprice, Rafi Hayne, and Dmitry Berenson. Predicting human reaching motion in collaborative tasks using inverse optimal control and iterative re-planning. In *2015 IEEE International Conference on Robotics and Automation (ICRA)*,

pages 885–892. IEEE, 2015.

[43] Ozgur S. Oguz, Zhehua Zhou, Stefan Glasauer, and Dirk Wollherr. An inverse optimal control approach to explain human arm reaching control based on multiple internal models. *Scientific Reports*, 8(1):5583, 2018.

[44] Anuj Karpatne, Gowtham Atluri, James H Faghmous, Michael Steinbach, Arindam Banerjee, Auroop Ganguly, Shashi Shekhar, Nagiza Samatova, and Vipin Kumar. Theory-guided data science: A new paradigm for scientific discovery from data. *IEEE Transactions on knowledge and data engineering*, 29(10):2318–2331, 2017.

[45] Isabelle Gauger, Tobias Nagel, and Marco Huber. Hybrides maschinelles lernen im kontext der produktion. In *Digitalisierung souverän gestalten II*, pages 64–79. Springer, 2022.

[46] Mauricio A. Álvarez, David Luengo, and Neil D. Lawrence. Latent force models. In *International Conference on Artificial Intelligence and Statistics*, pages 9–16, 2009.

[47] Micha Hersch, Florent Guenter, Sylvain Calinon, and Aude Billard. Dynamical system modulation for robot learning via kinesthetic demonstrations. *IEEE Transactions on Robotics*, 24(6):1463–1467, 2008.

[48] Auke Jan Ijspeert, Jun Nakanishi, Heiko Hoffmann, Peter Pastor, and Stefan Schaal. Dynamical movement primitives: Learning attractor models for motor behaviors. *Neural Computation*, 25(2):328–373, February 2013.

[49] Affan Pervez, Yuecheng Mao, and Dongheui Lee. Learning deep movement primitives using convolutional neural networks. In *2017 IEEE-RAS 17th International Conference on Humanoid Robotics (Humanoids)*, pages 191–197, 2017.

[50] Dorothea Koert, Joni Pajarinen, Albert Schotschneider, Susanne Trick, Constantin Rothkopf, and Jan Peters. Learning intention aware online adaptation of movement primitives. *IEEE Robotics and Automation Letters*, 4(4):3719–3726, 2019.

[51] Mahdi Khoramshahi and Aude Billard. A dynamical system approach to task-adaptation in physical human–robot interaction. *Autonomous Robots*, 43(4):927–946, 2019.

[52] Bruno Siciliano, Lorenzo Sciavicco, Luigi Villani, and Giuseppe Oriolo. *Robotics: modelling, planning and control*. Springer Science & Business Media, 2010.

[53] Emanuele Magrini and Alessandro De Luca. Hybrid force/velocity control for phys-

ical human-robot collaboration tasks. In *2016 IEEE/RSJ International Conference on Intelligent Robots and Systems (IROS)*, pages 857–863, 2016.

[54] Neville Hogan. Impedance Control: An Approach to Manipulation: Part I-Theory. *Journal of Dynamic Systems, Measurement, and Control*, 107(1):1–7, 03 1985.

[55] Fares J. Abu-Dakka and Matteo Saveriano. Variable impedance control and learning - a review. *Frontiers in Robotics and AI*, 7:177, 2020.

[56] Mojtaba Sharifi, Amir Zakerimanesh, Javad K. Mehr, Ali Torabi, Vivian K. Mushahwar, and Mahdi Tavakoli. Impedance variation and learning strategies in human-robot interaction. *IEEE Transactions on Cybernetics*, pages 1–14, 2021.

[57] Fanny Ficuciello, Luigi Villani, and Bruno Siciliano. Variable impedance control of redundant manipulators for intuitive human-robot physical interaction. *IEEE Transactions on Robotics*, 31(4):850–863, 2015.

[58] Vincent Duchaine and Clément Gosselin. Safe, stable and intuitive control for physical human-robot interaction. In *2009 IEEE International Conference on Robotics and Automation*, pages 3383–3388. IEEE, 2009.

[59] L. Peternel, N. Tsagarakis, and A. Ajoudani. Towards multi-modal intention interfaces for human-robot co-manipulation. *2016 IEEE/RSJ International Conference on Intelligent Robots and Systems (IROS)*, pages 2663–2669, 2016.

[60] Loris Roveda, Niccolò Iannacci, Federico Vicentini, Nicola Pedrocchi, Francesco Braghin, and Lorenzo Molinari Tosatti. Optimal impedance force-tracking control design with impact formulation for interaction tasks. *IEEE Robotics and Automation Letters*, 1(1):130–136, 2016.

[61] Theodoros Stouraitis, Lei Yan, João Moura, Michael Gienger, and Sethu Vijayakumar. Multi-mode trajectory optimization for impact-aware manipulation. In *2020 IEEE/RSJ International Conference on Intelligent Robots and Systems (IROS)*, pages 9425–9432, 2020.

[62] Maciej Bednarczyk, Hassan Omran, and Bernard Bayle. Model predictive impedance control. In *2020 IEEE International Conference on Robotics and Automation (ICRA)*, pages 4702–4708, 2020.

[63] Sami Haddadin and Elizabeth Croft. *Physical Human–Robot Interaction*, pages 1835–1874. Springer International Publishing, Cham, 2016.

[64] Przemyslaw A. Lasota. *Robust human motion prediction for safe and efficient human-robot interaction*. PhD thesis, Massachusetts Institute of Technology, 2019.

[65] Claudia Pérez D'Arpino. *Hybrid learning for multi-step manipulation in collaborative robotics.* PhD thesis, Massachusetts Institute of Technology, 2019.

[66] Changliu Liu. *Designing robot behavior in human-robot interactions.* PhD thesis, University of California, Berkeley, 2017.

[67] Medina Hernandez and José L. Ramon. *Model-based Control and Learning in Physical Human-Robot Interaction.* PhD thesis, Technical University of Munich, 2015.

[68] Carolyn Anglin and UP Wyss. Review of arm motion analyses. *Proceedings of the Institution of Mechanical Engineers, Part H: Journal of Engineering in Medicine,* 214(5):541–555, 2000.

[69] Damien Kelly and Frank Boland. Motion model selection in tracking humans. In *2006 IET Irish Signals and Systems Conference,* pages 363–368, 2006.

[70] Kevin P Murphy. *Machine learning: a probabilistic perspective.* MIT press, 2012.

[71] Markus Huber, Markus Rickert, Alois Knoll, Thomas Brandt, and Stefan Glasauer. Human-robot interaction in handing-over tasks. In *RO-MAN 2008 - The 17th IEEE International Symposium on Robot and Human Interactive Communication,* pages 107–112, 2008.

[72] Hai Zhu and Javier Alonso-Mora. Chance-constrained collision avoidance for mavs in dynamic environments. *IEEE Robotics and Automation Letters,* 4(2):776–783, 2019.

[73] Ugo Pattacini, Francesco Nori, Lorenzo Natale, Giorgio Metta, and Giulio Sandini. An experimental evaluation of a novel minimum-jerk cartesian controller for humanoid robots. In *2010 IEEE/RSJ International Conference on Intelligent Robots and Systems,* pages 1668–1674, 2010.

[74] Mahdi Ghazaei, Anders Robertsson, and Rolf Johansson. Online minimum-jerk trajectory generation. In *Proc. IMA Conf. Mathematics of Robotics,* 2015.

[75] Stephen H Scott. Optimal feedback control and the neural basis of volitional motor control. *Nature Reviews Neuroscience,* 5(7):532–545, 2004.

[76] Sergey Levine and Vladlen Koltun. Continuous inverse optimal control with locally optimal examples. In *Proceedings of the 29th International Coference on International Conference on Machine Learning,* pages 475–482, Madison, WI, USA, 2012. Omnipress.

[77] Bastien Berret, Enrico Chiovetto, Francesco Nori, and Thierry Pozzo. Evidence for

composite cost functions in arm movement planning: An inverse optimal control approach. *PLOS Computational Biology*, 7(10):1–18, 10 2011.

[78] Katja Mombaur, Anh Truong, and Jean-Paul Laumond. From human to humanoid locomotion–an inverse optimal control approach. *Autonomous Robots*, 28(3):369–383, 2010.

[79] Anne-Sophie Puydupin-Jamin, Miles Johnson, and Timothy Bretl. A convex approach to inverse optimal control and its application to modeling human locomotion. In *2012 IEEE International Conference on Robotics and Automation*, pages 531–536, 2012.

[80] Jairo Inga, Florian Köpf, Michael Flad, and Sören Hohmann. Individual human behavior identification using an inverse reinforcement learning method. In *2017 IEEE International Conference on Systems, Man, and Cybernetics (SMC)*, pages 99–104. IEEE, 2017.

[81] Sebastian Albrecht. *Modeling and numerical solution of inverse optimal control problems for the analysis of human motions*. PhD thesis, Technische Universität München, 2013.

[82] Jack M. Wang, David J. Fleet, and Aaron Hertzmann. Gaussian process dynamical models for human motion. *IEEE Transactions on Pattern Analysis and Machine Intelligence*, 30(2):283–298, 2008.

[83] Hildo Bijl, Thomas B. Schön, Jan-Willem van Wingerden, and Michel Verhaegen. System identification through online sparse gaussian process regression with input noise. *IFAC Journal of Systems and Control*, 2:1–11, 2017.

[84] Donald E Kirk. *Optimal control theory: an introduction*. Courier Corporation, 2004.

[85] Brian D Ziebart, Andrew L Maas, J Andrew Bagnell, Anind K Dey, et al. Maximum entropy inverse reinforcement learning. In *AAAI*, volume 8, pages 1433–1438. Chicago, IL, USA, 2008.

[86] Jim Mainprice, Rafi Hayne, and Dmitry Berenson. Goal set inverse optimal control and iterative replanning for predicting human reaching motions in shared workspaces. *IEEE Transactions on Robotics*, 32(4):897–908, 2016.

[87] Wanxin Jin, Dana Kulić, Shaoshuai Mou, and Sandra Hirche. Inverse optimal control from incomplete trajectory observations. *The International Journal of Robotics Research*, 40(6-7):848–865, 2021.

[88] Marcel Menner, Peter Worsnop, and Melanie N. Zeilinger. Constrained inverse optimal control with application to a human manipulation task. *IEEE Transactions on Control Systems Technology*, 29(2):826–834, 2021.

[89] Carl Edward Rasmussen and Christopher K. I. Williams. *Gaussian processes for machine learning*. Adaptive computation and machine learning. MIT Press, 2006.

[90] M. Huber. Recursive gaussian process: On-line regression and learning. *Pattern Recognit. Lett.*, 45:85–91, 2014.

[91] Andrew Mchutchon and Carl Rasmussen. Gaussian process training with input noise. In J. Shawe-Taylor, R. Zemel, P. Bartlett, F. Pereira, and K. Q. Weinberger, editors, *Advances in Neural Information Processing Systems*, volume 24. Curran Associates, Inc., 2011.

[92] Michalis K Titsias, Neil D Lawrence, and Magnus Rattray. Efficient sampling for gaussian process inference using control variables. In *NIPS*, pages 1681–1688. Citeseer, 2008.

[93] Mauricio Álvarez, David Luengo, Michalis Titsias, and Neil D Lawrence. Efficient multioutput gaussian processes through variational inducing kernels. In *Proceedings of the Thirteenth International Conference on Artificial Intelligence and Statistics*, pages 25–32. JMLR Workshop and Conference Proceedings, 2010.

[94] Mauricio Alvarez, Jan Peters, Neil Lawrence, and Bernhard Schölkopf. Switched latent force models for movement segmentation. *Advances in neural information processing systems*, 23:55–63, 2010.

[95] S. Mohammad Khansari-Zadeh and Aude Billard. Learning stable nonlinear dynamical systems with gaussian mixture models. *IEEE Transactions on Robotics*, 27(5):943–957, 2011.

[96] Mahdi Khoramshahi, Antoine Laurens, Thomas Triquet, and Aude Billard. From human physical interaction to online motion adaptation using parameterized dynamical systems. In *2018 IEEE/RSJ International Conference on Intelligent Robots and Systems (IROS)*, pages 1361–1366, 2018.

[97] Matteo Saveriano, Fares J Abu-Dakka, Aljaz Kramberger, and Luka Peternel. Dynamic movement primitives in robotics: A tutorial survey. *arXiv preprint arXiv:2102.03861*, 2021.

[98] Alexandros Paraschos, Christian Daniel, Jan Peters, Gerhard Neumann, et al. Probabilistic movement primitives. *Advances in neural information processing systems*, 2013.

[99] D. Park, H. Hoffmann, P. Pastor, and S. Schaal. Movement reproduction and obstacle avoidance with dynamic movement primitives and potential fields. In *Humanoids 2008 - 8th IEEE-RAS International Conference on Humanoid Robots*, pages 91–98, 2008.

[100] Andrej Gams, Tadej Petrič, Martin Do, Bojan Nemec, Jun Morimoto, Tamim Asfour, and Aleš Ude. Adaptation and coaching of periodic motion primitives through physical and visual interaction. *Robotics and Autonomous Systems*, 75:340–351, 2016.

[101] Leonidas Koutras and Zoe Doulgeri. Dynamic movement primitives for moving goals with temporal scaling adaptation. In *2020 IEEE International Conference on Robotics and Automation (ICRA)*, pages 144–150. IEEE, 2020.

[102] H. Ben Amor, G. Neumann, S. Kamthe, O. Kroemer, and J. Peters. Interaction primitives for human-robot cooperation tasks. In *2014 IEEE International Conference on Robotics and Automation (ICRA)*, pages 2831–2837, 2014.

[103] Guilherme Maeda, Gerhard Neumann, Marco Ewerton, Rudolf Lioutikov, Oliver Kroemer, and Jan Peters. Probabilistic movement primitives for coordination of multiple human-robot collaborative tasks. *Autonomous Robots*, 41, 03 2017.

[104] Adam Conkey and Tucker Hermans. Active learning of probabilistic movement primitives. In *2019 IEEE-RAS 19th International Conference on Humanoid Robots (Humanoids)*, pages 1–8, 2019.

[105] Guilherme Maeda, Okan Koç, and Jun Morimoto. Phase portraits as movement primitives for fast humanoid robot control. *Neural Networks*, 129:109–122, 2020.

[106] Jonas Umlauft, Dominik Sieber, and Sandra Hirche. Dynamic movement primitives for cooperative manipulation and synchronized motions. In *2014 IEEE International Conference on Robotics and Automation (ICRA)*, pages 766–771, 2014.

[107] Tomas Kulvicius, Martin Biehl, Mohamad Javad Aein, Minija Tamosiunaite, and Florentin Wörgötter. Interaction learning for dynamic movement primitives used in cooperative robotic tasks. *Robotics and Autonomous Systems*, 61(12):1450–1459, 2013.

[108] Y. Fanger, J. Umlauft, and S. Hirche. Gaussian processes for dynamic movement primitives with application in knowledge-based cooperation. In *2016 IEEE/RSJ International Conference on Intelligent Robots and Systems (IROS)*, pages 3913–3919, 2016.

[109] L. Koutras and Z. Doulgeri. A novel dmp formulation for global and frame in-

dependent spatial scaling in the task space. In *2020 29th IEEE International Conference on Robot and Human Interactive Communication (RO-MAN)*, pages 727–732, 2020.

[110] Miguel Prada, Anthony Remazeilles, Ansgar Koene, and Satoshi Endo. Dynamic movement primitives for human-robot interaction: Comparison with human behavioral observation. In *2013 IEEE/RSJ International Conference on Intelligent Robots and Systems*, pages 1168–1175, Tokyo, Japan, 2013. IEEE.

[111] D. Widmann and Y. Karayiannidis. Human motion prediction in human-robot handovers based on dynamic movement primitives. In *2018 European Control Conference (ECC)*, pages 2781–2787, 2018.

[112] A. Fishman, C. Paxton, W. Yang, D. Fox, B. Boots, and N. Ratliff. Collaborative interaction models for optimized human-robot teamwork. In *2020 IEEE/RSJ International Conference on Intelligent Robots and Systems (IROS)*, pages 11221–11228, 2020.

[113] Tie Wang, Goran S. Dordevic, and Reza Shadmehr. Learning the dynamics of reaching movements results in the modification of arm impedance and long-latency perturbation responses. *Biological Cybernetics*, 85(6):437–448, 2001.

[114] Xianghui Yuan, Chongzhao Han, Zhansheng Duan, and Ming Lei. Comparison and choice of models in tracking target with coordinated turn motion. In *2005 7th International Conference on Information Fusion*, volume 2, pages 6 pp.–, 2005.

[115] Przemyslaw A Lasota and Julie A Shah. A multiple-predictor approach to human motion prediction. In *2017 IEEE International Conference on Robotics and Automation (ICRA)*, pages 2300–2307. IEEE, 2017.

[116] Qinghua Li, Zhao Zhang, Yue You, Yaqi Mu, and Chao Feng. Data driven models for human motion prediction in human-robot collaboration. *IEEE Access*, 8:227690–227702, 2020.

[117] Vibekananda Dutta and Teresa Zielinska. Predicting human actions taking into account object affordances. *Journal of Intelligent and Robotic Systems*, 93(3):745–761, 2019.

[118] J.D. Morrow and P.K. Khosla. Manipulation task primitives for composing robot skills. In *Proceedings of International Conference on Robotics and Automation*, volume 4, pages 3354–3359 vol.4, 1997.

[119] Claudia Pérez-D'Arpino and Julie A Shah. C-learn: Learning geometric constraints from demonstrations for multi-step manipulation in shared autonomy. In *2017*

IEEE International Conference on Robotics and Automation (ICRA), pages 4058–4065. IEEE, 2017.

[120] Peter Pastor, Mrinal Kalakrishnan, Franziska Meier, Freek Stulp, Jonas Buchli, Evangelos Theodorou, and Stefan Schaal. From dynamic movement primitives to associative skill memories. *Robotics and Autonomous Systems*, 61(4):351–361, 2013.

[121] Junjun Li, Zhijun Li, Xinde Li, Ying Feng, Yingbai Hu, and Bugong Xu. Skill learning strategy based on dynamic motion primitives for human–robot cooperative manipulation. *IEEE Transactions on Cognitive and Developmental Systems*, 13(1):105–117, 2020.

[122] Mrinal Kalakrishnan, Sachin Chitta, Evangelos Theodorou, Peter Pastor, and Stefan Schaal. Stomp: Stochastic trajectory optimization for motion planning. In *2011 IEEE international conference on robotics and automation*, pages 4569–4574. IEEE, 2011.

[123] Freek Stulp, Jonathan Grizou, Baptiste Busch, and Manuel Lopes. Facilitating intention prediction for humans by optimizing robot motions. In *2015 IEEE/RSJ International Conference on Intelligent Robots and Systems (IROS)*, pages 1249–1255, 2015.

[124] Haitao Liu, Yew-Soon Ong, Xiaobo Shen, and Jianfei Cai. When gaussian process meets big data: A review of scalable gps. *IEEE Transactions on Neural Networks and Learning Systems*, 31(11):4405–4423, 2020.

[125] Claudio Santos Pinhanez, Heloisa Candello, Paulo Cavalin, Mauro Carlos Pichiliani, Ana Paula Appel, Victor Henrique Alves Ribeiro, Julio Nogima, Maira de Bayser, Melina Guerra, Henrique Ferreira, et al. Integrating machine learning data with symbolic knowledge from collaboration practices of curators to improve conversational systems. In *Proceedings of the 2021 CHI Conference on Human Factors in Computing Systems*, pages 1–13, 2021.

[126] Muhammad Awais and Dominik Henrich. Human-robot collaboration by intention recognition using probabilistic state machines. In *19th International Workshop on Robotics in Alpe-Adria-Danube Region (RAAD 2010)*, pages 75–80, 2010.

[127] Stefanos Nikolaidis, Swaprava Nath, Ariel D Procaccia, and Siddhartha Srinivasa. Game-theoretic modeling of human adaptation in human-robot collaboration. In *Proceedings of the 2017 ACM/IEEE international conference on human-robot interaction*, pages 323–331, 2017.

[128] Sebastian Erhart. *Cooperative multi-robot manipulation under uncertain kinematic grasp parameters.* Dissertation, Technische Universität München, München, 2016.

[129] Nathanaël Jarrassé, Themistoklis Charalambous, and Etienne Burdet. A framework to describe, analyze and generate interactive motor behaviors. *PloS one*, 7(11):e49945, 2012.

[130] Luka Peternel and Jan Babic. Learning of compliant human-robot interaction using full-body haptic interface. *Advanced Robotics*, 27, 09 2013.

[131] Paul C Watson. Instrumented remote center compliance device, February 23 1982. US Patent 4,316,329.

[132] Christian Ott. *Cartesian impedance control of redundant and flexible-joint robots.* Springer, 2008.

[133] Sebastian Erhart and Sandra Hirche. Model and analysis of the interaction dynamics in cooperative manipulation tasks. *IEEE Transactions on Robotics*, 32(3):672–683, 2016.

[134] D. Williams and O. Khatib. The virtual linkage: a model for internal forces in multi-grasp manipulation. In *[1993] Proceedings IEEE International Conference on Robotics and Automation*, pages 1025–1030 vol.1, 1993.

[135] Denis Cehajic, Sebastian Erhart, and Sandra Hirche. Grasp pose estimation in human-robot manipulation tasks using wearable motion sensors. In *2015 IEEE/RSJ International Conference on Intelligent Robots and Systems (IROS)*, pages 1031–1036, 2015.

[136] Etienne Burdet, Rieko Osu, David W. Franklin, Theodore E. Milner, and Mitsuo Kawato. The central nervous system stabilizes unstable dynamics by learning optimal impedance. *Nature*, 414(6862):446–449, 2001.

[137] Yanan Li and Shuzhi Sam Ge. Human-robot collaboration based on motion intention estimation. *IEEE/ASME Transactions on Mechatronics*, 19(3):1007–1014, 2014.

[138] José R. Medina, Hendrik Börner, Satoshi Endo, and Sandra Hirche. Impedance-based gaussian processes for modeling human motor behavior in physical and non-physical interaction. *IEEE Transactions on Biomedical Engineering*, 66(9):2499–2511, 2019.

[139] Hendrik Börner, Satoshi Endo, and Sandra Hirche. Estimation of involuntary components of human arm impedance in multi-joint movements via feedback jerk

isolation. *Frontiers in Neuroscience*, 14:459, 2020.

[140] Yanan Li, Keng Peng Tee, Rui Yan, Wei Liang Chan, and Yan Wu. A framework of human–robot coordination based on game theory and policy iteration. *IEEE Transactions on Robotics*, 32(6):1408–1418, 2016.

[141] Frank L Lewis, Draguna Vrabie, and Vassilis L Syrmos. *Optimal control*. John Wiley & Sons, 2012.

[142] Yanan Li, Gerolamo Carboni, Franck Gonzalez, Domenico Campolo, and Etienne Burdet. Differential game theory for versatile physical human–robot interaction. *Nature Machine Intelligence*, 1(1):36–43, 2019.

[143] Kyriakos G Vamvoudakis and Frank L Lewis. Multi-player non-zero-sum games: Online adaptive learning solution of coupled hamilton–jacobi equations. *Automatica*, 47(8):1556–1569, 2011.

[144] C. Yang, G. Ganesh, S. Haddadin, S. Parusel, A. Albu-Schaeffer, and E. Burdet. Human-like adaptation of force and impedance in stable and unstable interactions. *IEEE Transactions on Robotics*, 27(5):918–930, Oct 2011.

[145] Jonas Buchli, Freek Stulp, Evangelos Theodorou, and Stefan Schaal. Learning variable impedance control. *The International Journal of Robotics Research*, 30(7):820–833, 2011.

[146] H. Modares, I. Ranatunga, F. L. Lewis, and D. O. Popa. Optimized assistive human-robot interaction using reinforcement learning. *IEEE Transactions on Cybernetics*, 46(3):655–667, March 2016.

[147] F. Dimeas and N. Aspragathos. Reinforcement learning of variable admittance control for human-robot co-manipulation. In *2015 IEEE/RSJ International Conference on Intelligent Robots and Systems (IROS)*, pages 1011–1016, Sep. 2015.

[148] Richard S Sutton and Andrew G Barto. *Reinforcement learning: An introduction*. MIT press, 2018.

[149] Leslie Pack Kaelbling, Michael L Littman, and Andrew W Moore. Reinforcement learning: A survey. *Journal of artificial intelligence research*, 4:237–285, 1996.

[150] Richard Bellman. Dynamic programming. *Science*, 153(3731):34–37, 1966.

[151] Jens Kober, J Andrew Bagnell, and Jan Peters. Reinforcement learning in robotics: A survey. *The International Journal of Robotics Research*, 32(11):1238–1274, 2013.

[152] Athanasios S Polydoros and Lazaros Nalpantidis. Survey of model-based reinforce-

ment learning: Applications on robotics. *Journal of Intelligent & Robotic Systems*, 86(2):153–173, 2017.

[153] Daniel Görges. Relations between model predictive control and reinforcement learning. *IFAC-PapersOnLine*, 50(1):4920–4928, 2017.

[154] Said G. Khan, Guido Herrmann, Frank L. Lewis, Tony Pipe, and Chris Melhuish. Reinforcement learning and optimal adaptive control: An overview and implementation examples. *Annual Reviews in Control*, 36(1):42 – 59, 2012.

[155] Daniel Görges. Distributed adaptive linear quadratic control using distributed reinforcement learning. *IFAC-PapersOnLine*, 52(11):218–223, 2019.

[156] Feng Wan, Li-Xin Wang, He-Yun Zhu, and You-Xian Sun. Generating persistently exciting inputs for nonlinear dynamic system identification using fuzzy models. In *10th IEEE International Conference on Fuzzy Systems.(Cat. No. 01CH37297)*, volume 1, pages 505–508. IEEE, 2001.

[157] Steven J Bradtke, B Erik Ydstie, and Andrew G Barto. Adaptive linear quadratic control using policy iteration. In *Proceedings of 1994 American Control Conference-ACC'94*, volume 3, pages 3475–3479. IEEE, 1994.

[158] Zaiwei Chen, Sheng Zhang, Thinh T Doan, Siva Theja Maguluri, and John-Paul Clarke. Performance of q-learning with linear function approximation: Stability and finite-time analysis. *arXiv preprint arXiv:1905.11425*, 2019.

[159] S. N. Balakrishnan, Jie Ding, and Frank L. Lewis. Issues on stability of adp feedback controllers for dynamical systems. *IEEE Transactions on Systems, Man, and Cybernetics, Part B (Cybernetics)*, 38(4):913–917, 2008.

[160] Yongqiang Li, Chengzan Yang, Zhongsheng Hou, Yuanjing Feng, and Chenkun Yin. Data-driven approximate q-learning stabilization with optimality error bound analysis. *Automatica*, 103:435–442, 2019.

[161] A. Kucukyilmaz, T. M. Sezgin, and C. Basdogan. Conveying intentions through haptics in human-computer collaboration. In *2011 IEEE World Haptics Conference*, pages 421–426, June 2011.

[162] Jonas Schmidtler and Klaus Bengler. Fast or accurate? -performance measurements for physical human-robot collaborations. *Procedia Manufacturing*, 3:1387 – 1394, 2015. 6th International Conference on Applied Human Factors and Ergonomics (AHFE 2015) and the Affiliated Conferences, AHFE 2015.

[163] Arash Ajoudani, Nikos G. Tsagarakis, and Antonio Bicchi. Choosing poses for force

and stiffness control. *IEEE Transactions on Robotics*, 33(6):1483–1490, 2017.

[164] Nathanaäl Jarrassé, Vittorio Sanguineti, and Etienne Burdet. Slaves no longer: review on role assignment for human-robot joint motor action. *Adaptive Behavior*, 22(1):70–82, 2014.

[165] Alexander Mörtl, Martin Lawitzky, Ayse Kucukyilmaz, Metin Sezgin, Cagatay Basdogan, and Sandra Hirche. The role of roles: Physical cooperation between humans and robots. *The International Journal of Robotics Research*, 31(13):1656–1674, 2012.

[166] Yanan Li, Keng Peng Tee, Wei Liang Chan, Rui Yan, Yuanwei Chua, and Dilip Kumar Limbu. Continuous role adaptation for human–robot shared control. *IEEE Transactions on Robotics*, 31(3):672–681, 2015.

[167] Bryan Whitsell and Panagiotis Artemiadis. On the role duality and switching in human-robot cooperation: An adaptive approach. In *2015 IEEE International Conference on Robotics and Automation (ICRA)*, pages 3770–3775. IEEE, 2015.

[168] Vladislav Golyanik, Bertram Taetz, and Didier Stricker. Joint pre-alignment and robust rigid point set registration. pages 4503–4507, Phoenix, AZ, USA, 2016. IEEE.

[169] Z. Li and Kris K. Hauser. Predicting object transfer position and timing in human-robot handover tasks. 2015.

[170] Guy Hoffman. Evaluating fluency in human robot collaboration. *IEEE Transactions on Human-Machine Systems*, 49:209–218, 2019.

[171] Sivakumar Balasubramanian, Alejandro Melendez-Calderon, and Etienne Burdet. A robust and sensitive metric for quantifying movement smoothness. *IEEE Transactions on Biomedical Engineering*, 59:2126–2136, 2012.

[172] Anca D. Dragan, Kenton C.T. Lee, and Siddhartha S. Srinivasa. Legibility and predictability of robot motion. pages 301–308, Tokyo, Japan, 2013. IEEE.

[173] Clint Hansen, Paula Arambel, Khalil Ben Mansour, Veronique Perdereau, and Frdéric Marin. Human-human handover tasks and how distance and object mass matter. *Perceptual and Motor Skills*, 124(1):182–199, 2017. PMID: 30208781.

[174] Jimmy Baraglia, Maya Cakmak, Yukie Nagai, Rajesh PN Rao, and Minoru Asada. Efficient human-robot collaboration: when should a robot take initiative? *The International Journal of Robotics Research*, 36(5-7):563–579, 2017.

[175] Yaakov Engel, Shie Mannor, and Ron Meir. Reinforcement learning with gaussian

processes. In *Proceedings of the 22nd international conference on Machine learning*, pages 201–208, 2005.

[176] Nadia Barbara Figueroa Fernandez and Aude Billard. Modeling compositions of impedance-based primitives via dynamical systems. In *Proceedings of the Workshop on Cognitive Whole-Body Control for Compliant Robot Manipulation (COWB-COMP)*, number CONF, 2018.

Zusammenfassung

In den letzten Jahren wurden immer mehr kollaborative Roboter (kurz "Cobot") nicht nur in modernen Produktionssystemen, sondern auch in alltäglichen persönlichen Dienstleistungen wie der Krankenpflege, dem Haushalt und der Physiotherapie, eingesetzt. Trotz der rasanten Entwicklung der Technologien und der Forschung in diesem Bereich stehen die kollaborative Robotik heute noch vor großen Herausforderungen. Die Technik und die akademischen Entwicklungen haben große Erfolge bei der Lösung des Problems der "Koexistenz" zwischen Menschen und Robotern in unmittelbarer Nähe erzielt. Allerdings ist es nicht nur notwendig für den Roboter, die Sicherheit des Menschen zu gewährleisten, sondern auch die Absicht des Menschen anzuerkennen, einen proaktiven Beitrag für die gesamte Aufgabe zu leisten und mit dem Menschen zu interagieren.

Ziel dieser Arbeit ist die Entwicklung eines methodisch neuen Ansatzes für die Bewegungsprädiktion und Interaktionsregelung in Mensch-Roboter-Kollaboration. Die Forschungsfragen und Beiträge der Arbeit werden folgendermaßen erläutert.

Echtzeitfähige Bewegungsprädiktion

Die Modellierung und Prädiktion menschlicher Bewegungen unter Berücksichtigung von Genauigkeit und Berechnungseffizienz ist eine Herausforderung. Anstatt präziser, aber arbeitsintensiver biomechanischer Modelle zielt diese Arbeit darauf ab, einen Modellierungsansatz zu finden, der die wichtigsten menschlichen Bewegungsmerkmale abdeckt, wenig Sensorinformationen benötigt und sich online angepasst werden kann. Zu diesem Zweck werden in Kapitel 2 zunächst die klassischen rekursiven Bayes'sche Schätzungsansätze mittels linearer, physikalischer Kinematik-Modelle untersucht. Zwei Konzepte, der Kalman-Filter mit konstantem Beschleunigungsmodell und der rekursive Least-Square-Filter mit Minimum-Jerk-Modell, werden an einem Datensatz von menschlichen Handbewegungen getestet. Die Ergebnisse zeigen, dass sie für kurzfristige Prädiktion mit weniger Unsicherheiten in einer statischen Umgebung geeignet sind. Der Fehler wächst jedoch proportional zu den Prädiktionsschritten, was ihre Begrenzung bei langfristiger Prädiktion zeigt.

Selbst für eine einfache Handbewegung, bei der nur translatorische Freiheitsgrade berücksichtigt werden, sind lineare Modelle mit additivem Gaußschen Rauschen nur in einem kleinen Bereich um eine bestimmte Geschwindigkeit gültig. Als nächsten Entwicklungsschritt werden in Kapitel 3 datengestützte Modellierungsansätze, die auf maschinellem Lernen beruhen, untersucht. Motiviert durch die Erkenntnisse aus der Forschung zur menschlichen Motorsteuerung, d.h., das menschliche Zentralnervensystem arbeitet als optimales Regelungssystem, konzentriert sich die erste Studie auf die Identifizierung der Zielfunktion, die der Mensch zu optimieren versucht. Eingesetzt wird das Verfahren zur inversen optimalen Regelungen basierend auf dem Prinzips der maximalen Entropie. Die Methode erfordert die Lösung von zwei eingeschränkten Optimierungsproblemen: eines für die Bestimmung der Kostenfunktion und das andere für die Reproduktion der menschlichen Trajektorien. Daher ist die Komplexität der Implementierung hoch. Die Ergebnisse zeigen, dass dieser Ansatz die Performance für langfristige Prädiktion verbessert, jedoch nicht signifikant. Die Online-Anpassungsfähigkeit ist gering, da die Methode in der Regel eine Betrachtung der vollständige Trajektorie erfordert.

Bei allen oben genannten Methoden muss man die Klasse der Funktionen einschränken, die zur Modellierung der Ein- und Ausgangsbeziehungen verwendet werden (z. B. lineare Funktionen). Wenn die gewählte Klasse von Funktionen das System nicht gut modellieren kann, wird die Prädiktion ungenau sein. Um dieses Problem zu überwinden, konzentriert sich der zweite Teil von Kapitel 3 auf nicht-parametrische Ansätze. Es wird ein rekursiver, spärlicher GP-Regressionsalgorithmus angewandt. Im Vergleich zur Standard-GP reduziert diese Methode die Berechnungskomplexität und macht das Modell anpassungsfähig an die eingehenden Daten. Die Ergebnisse zeigen, dass diese Methode sowohl bei kurzfristigen als auch bei langfristigen Prädiktionen gut funktioniert. Außerdem ist der rekursive Algorithmus in der Lage, mit neuen Daten aus unbekannten Bewegungsarten umzugehen, die nicht im Trainingssatz enthalten sind. Dennoch sind mehrere Iterationen erforderlich, um die Verteilung der neuen Daten zu lernen. Während dieses Prozesses sind die Vorhersagen nicht zuverlässig.

Nach der Untersuchung sowohl physikalischer als auch datengestützte Ansätze kommt man zu dem Schluss, dass sich ihre Eigenschaften in gewisser Weise ergänzen. Einfache physikalische Modelle lassen sich in der Regel in ihrem Gültigkeitsbereich gut verallgemeinern, erfordern keine Trainingsdaten, können aber nur einen begrenzten Teil des Systemverhaltens darstellen. Daten-basierte Ansätze haben eine starke Fähigkeit, komplexe Systeme darzustellen, sind sich der Unsicherheit bewusst, können aber nur in dem Bereich arbeiten, wo Daten verfügbar sind. Daher scheint ein hybrider physikalischer und daten-basierter Ansatz sehr vielversprechend. Dynamische Bewegungsprimitive gehören zu den am ausführlichsten untersuchten hybriden Modellierungsansätzen, die eine physikalisch gut verstandene lineare Attraktor-Dynamik um einen lernbaren nichtlinearen Zwangsterm erweitern. In Kapitel 4 wird die konventionelle DMP-Formulierung um einen GP-basierten nichtlinearen Term erweitert, um (1) dessen Fähigkeit zur Beschreibung nichtlinearer Dynamik zu nutzen und (2) den Aufwand für die manuelle Parametereinstellung zu reduzieren. Um mehrere praktische Probleme zu lösen, insbesondere die Empfindlichkeit des DMP-Modells gegenüber der variierenden Zielposition, wird ein Gewichtungsfaktor zur Anpassung der Dominanz linearer und nichtlinearer Terme und ein rotationsbasierter räumlicher Skalierungsfaktor entwickelt. Die Ergebnisse zeigen, dass dieser Ansatz alle anderen Methoden, die in dieser Arbeit implementiert wurden, übertrifft.

Kapitel 5 fasst alle in dieser Arbeit untersuchten Methoden zur Prädiktion menschlicher Bewegungen zusammen und erörtert ausführlich ihren Anwendungsbereich, ihre Genauigkeit in verschiedenen Zeitskalen und ihre Implementierungskomplexität. Anschließend wird ein Ausblick auf mögliche zukünftige Forschungstrends in der Bewegungsprädiktion mittels hybrider Ansätze gegeben.

Adaptive Interaktionsregelung

Die Prädiktion menschlicher Bewegungen hilft bei der Erstellung menschengerechter Robotertrajektorien. Der zweite Teil der Arbeit befasst sich mit der Ausführungsphase. Wenn während der Zusammenarbeit kein physischer Kontakt stattfindet, muss nur die Trajektorienverfolgung des Roboters berücksichtigt werden. Viele Kollaborationsaufgaben erfordern jedoch physische Interaktionen zwischen Menschen und Robotern, die eine größere Herausforderung für den Regelungsentwurf darstellen. Neben der Sicherheit sollten Roboter in der Lage sein, proaktive Beiträge zu leisten, um den physischen und koordinativen Aufwand für den Menschen weiter zu reduzieren.

Zu diesem Zweck wurde in Kapitel 6 das dynamische Verhalten des gesamten Systems untersucht, indem Menschen und Roboter gemeinsam ein starres Objekt übertragen. Anstelle einer umfassenden kinematischen und dynamischen Modellierung jedes einzelnen Agenten unter Berücksichtigung aller Freiheitsgrade besteht ein effizienterer Ansatz darin, das nachgiebige Regelungsverhalten sowohl von Robotern als auch von Menschen durch physikalische Randbedingungen in die Objektdynamik einzubeziehen. Die Impedanzregelung ist ein bekannter Ansatz zur Nachgiebigkeitsregelung, bei dem sowohl Menschen als auch Roboter als ein Masse-Feder-Dämpfer-System zweiter Ordnung betrachtet werden können. Aus der Sicht der Modellierung können daher physische Mensch-Roboter-Kollaborationsaufgaben näherungsweise durch eine ernsthafte Kopplung mechanischer Impedanzen beschrieben werden, was die Modellkomplexität erheblich vereinfacht.

Die Koordinierung zwischen Menschen und Robotern kann durch Differenzialspiele formuliert werden, bei denen alle Teilnehmer versuchen, ein Optimierungsproblem kooperativ zu lösen. Es ist zu beachten, dass es selbst für das vereinfachte Modell schwierig ist, ein Regelungsverfahren mit einem expliziten modellbasierten Ansatz zu entwerfen, da die Absichten von Menschen unbekannt und zeitlich variabel sind. Um dieses Problem zu lösen wird in Kapitel 7 erneut maschinelles Lernen mit physikalisch gut verstanden Regelungsmethoden kombiniert. Das Ergebnis ist eine adaptive Impedanzregelung, die auf bestärkendem Lernen basiert. Die Schlüsselidee besteht darin, die Reglerparameter während der Interaktion mit dem Menschen mithilfe des Q-Learning-Algorithmus anzupassen. Darüber hinaus werden auch Konzepte der Beschränkung der Kontaktkraft und der Handhabung von physikalischen Einschränkungen diskutiert.

Experimentelle Validierung

In Kapitel 8 werden die entwickelten Lern- und Regelungsmethoden durch mehrere Mensch-Roboter- Kollaborationsaufgaben auf einer 7-DOF-Roboterplattform validiert. Die Experimente umfassen zwei typische Benchmark-Anwendungen, nämlich Objektübergabe und Objekthandhabung, sodass sowohl kontaktlose als auch physische Interaktionen einbezogen werden.

Die Ergebnisse zeigen folgenden Fähigkeiten sowie Vorteile des entwickelten Systems:

- menschliche Bewegungsmodelle mit einer geringen Menge an Trainingsdaten zu erlernen,

- eine zuverlässige Online- Prädiktion menschlicher Bewegungen und eine frühzeitige Reaktion auf menschliche Bewegungsvariationen zu realisieren,

- proaktive Beiträge zu physischen Kollaborationsaufgaben zu liefern und sich nachgiebig in Reaktion auf Kontaktkräfte zu verhalten,

- sowohl die Performanzen (Erfolgsrate, Ausführungszeit, Positionsfehler) als auch die Benutzererfahrung (Komfort, Flüssigkeit) zu verbessern.

Im Kapitel 9 werden die Ergebnisse dieser Arbeit zusammengefasst und diskutiert. Zusätzlich wird ein Ausblick über mögliche zukünftige Forschungsfragen gegeben.

Supervised theses

1. Mirjan Heubaum. *Auslegung eines NTSM-Reglers für einen 2-DOF-Manipulator und Anwendung auf einem Franka Emika Panda Roboter*, Master project, 2021.

2. Shruti Shrikant Tayade. *Development of Simulation Models and Implementation of an Operation and Control Strategy for the Electrified Drivetrain of an Off-road Application Using the Example of a Mobile Excavator*, Master thesis, 2019.

3. Min Wang. *On-line Trajectory Planning based on Chance-Constraint Model Predictive Control and Gaussian Process Regression with an Application to Human-Robot Collaboration*, Master thesis, 2019.

4. Jia Lyu. *Schätzung der Kontaktstreifigkeit und Dämpfung für einen Roboterarm*, Bachelor thesis, 2018.

5. Dong Yang. *Untersuchungen zu kooperativen Manipulationsaufgaben mit ereignisbasierter Kommunikation*, Bachelor thesis, 2018.

6. Frank Staller. *Integration eines IMU Sensors in den Regelkreis eines mobilen Roboters*, Bachelor thesis, 2018.

7. Wenqian Xu. *Überarbeitung des Steuerungssystems für einen seriellen Roboterarm*, Bachelor thesis, 2017.

8. Xiang Chen. *Interaction Control of a 5-DOF Robot Arm with Cntact Force Estimation*, Master thesis, 2017.

9. Zhijie Lin. *Identification of Dynamic Parameters of a 5-DOF Robot Arm*, Master thesis, 2016.

Curriculum Vitae

Personal Data

Name	Min Wu
Born	08 August 1988, in Anhui ,China
E-Mail	mwu@eit.uni-kl.de

Education

10/2012 - 07/2015	**TU Kaiserslautern**
	Master Degree in Electrical and Computer Engineering Thesis: "Investigation on Collaboration of Multiple Mobile Manipulators"
03/2009 - 12/2012	**TU Kaiserslautern**
	Bachelor Degree in Electrical and Computer Engineering Thesis: "Statische und dynamische FEM-Berechnung eines elektromagnetischen Aktuators"
09/2006 - 03/2009	**Fuzhou University, China**
	Bachelor Degree in Electrical Engineering and Automation

Professional Experience

09/2015 -10/2020   **TU Kaiserslautern, Institute of Control Systems**

Research and Teaching Associate

03/2013 - 05/2015   **TU Kaiserslautern, Institute of Control Systems**

Student Research Assistant

10/2011 - 03/2012   **ZF Lenksysteme GmbH**

Internship

In der Reihe „*Forschungsberichte aus dem Lehrstuhl für Regelungssysteme*", herausgegeben von Steven Liu, sind bisher erschienen:

17	Hengyi Wang	Delta-connected Cascaded H-bridge Multilevel Converter as Shunt Active Power Filter	
		ISBN 978-3-8325-5015-8, 2019, 173 S.	38.00 €
18	Sebastian Caba	Energieoptimaler Betrieb gekoppelter Mehrpumpensysteme	
		ISBN 978-3-8325-5079-0, 2020, 141 S.	37.00 €
19	Alen Turnwald	Modelling and Control of an Autonomous Two-Wheeled Vehicle	
		ISBN 978-3-8325-5205-3, 2020, 175 S.	41.00 €
20	Zhuoqi Zeng	Ultra-wideband Based Indoor Localization Using Sensor Fusion and Support Vector Machine	
		ISBN 978-3-8325-5229-9, 2021, 152 S.	51.00 €
21	Yanhao He	Distributed Optimisation for Multi-Robot Cooperative Manipulation Control in Dynamic Environments	
		ISBN 978-3-8325-5440-8, 2022, 180 S.	48.50 €
22	Min Wu	A Hybrid Physical and Data-driven Approach to Motion Prediction and Control in Human-Robot Collaboration	
		ISBN 978-3-8325-5484-2, 2022, 212 S.	50.50 €

Alle erschienenen Bücher können unter der angegebenen ISBN im Buchhandel oder direkt beim Logos Verlag Berlin (www.logos-verlag.de, Fax: 030 - 42 85 10 92) bestellt werden.